HAWKER HURRICANE
THE HISTORY OF A LEGEND

WRITTEN BY MIKE LEPINE

sona BOOKS

Sona
BOOKS

© Sona Books Ltd 2024

First Published Sona Books Ltd 2024
© 2024 Danann media publishing Ltd
© Design 2024 Danann Media Publishing ltd

WARNING: For private domestic use only, any unauthorised copying, hiring, lending or public performance of this book is illegal.

CAT NO: SON0599

Photography courtesy of

Getty images, Alamy and Wiki Commons

Book layout & design: Darren Grice and Kevin Gardner

Proofreader: Finn O'Neill

All rights reserved. No Part of this title may be reproduced or transmitted in any material form (including photocopying or storing it in any medium by electronic means and whether or not transiently or incidentally to some other use of this publication) without the written permission of the copyright owner, except in accordance with the provisions of the Copyright, Designs and Patents Act 1988. Applications for the copyright owner's written permission should be addressed to the publisher.

Made in EU.
ISBN: 978-1-915343-55-0

CONTENTS

- THE BIRTH OF AN ALMOST-LEGEND — 10
- REAL WAR/PHONEY WAR — 28
- THE BATTLE OF BRITAIN — 46
- INTO THE BLITZ — 68
- SIGNIFICANT VERSIONS AND VARIANTS — 76
- FAR-FLUNG FIELDS — 88
- WAR IN THE MED — 94
- FRIENDS AND ENEMIES — 108
- 1942 — 116
- TIPPING THE BALANCE — 130
- THE LAST OF THE MANY — 136

'The Hurricane was a jolly good machine. A rugged type, stronger than the Spitfire.'
Hugh Dowding

'The Spitfire…was a Rolls-Royce. The Hurricane was a tank.'
Roger Wilkinson, Hurricane pilot, No 3 Squadron

'One hears so much about the superiority of the Spitfire over the Hurricane. It is all rubbish.'
Roland Beamont, Hurricane pilot and legendary test flier

'The Hurricane was hopeless… a nice plane to shoot down.'
Adolf Gallans

'(Hurricane pilots) seem to be forgotten, sir. It seems nobody loves us.'
Peter Wykeham, 43 Squadron

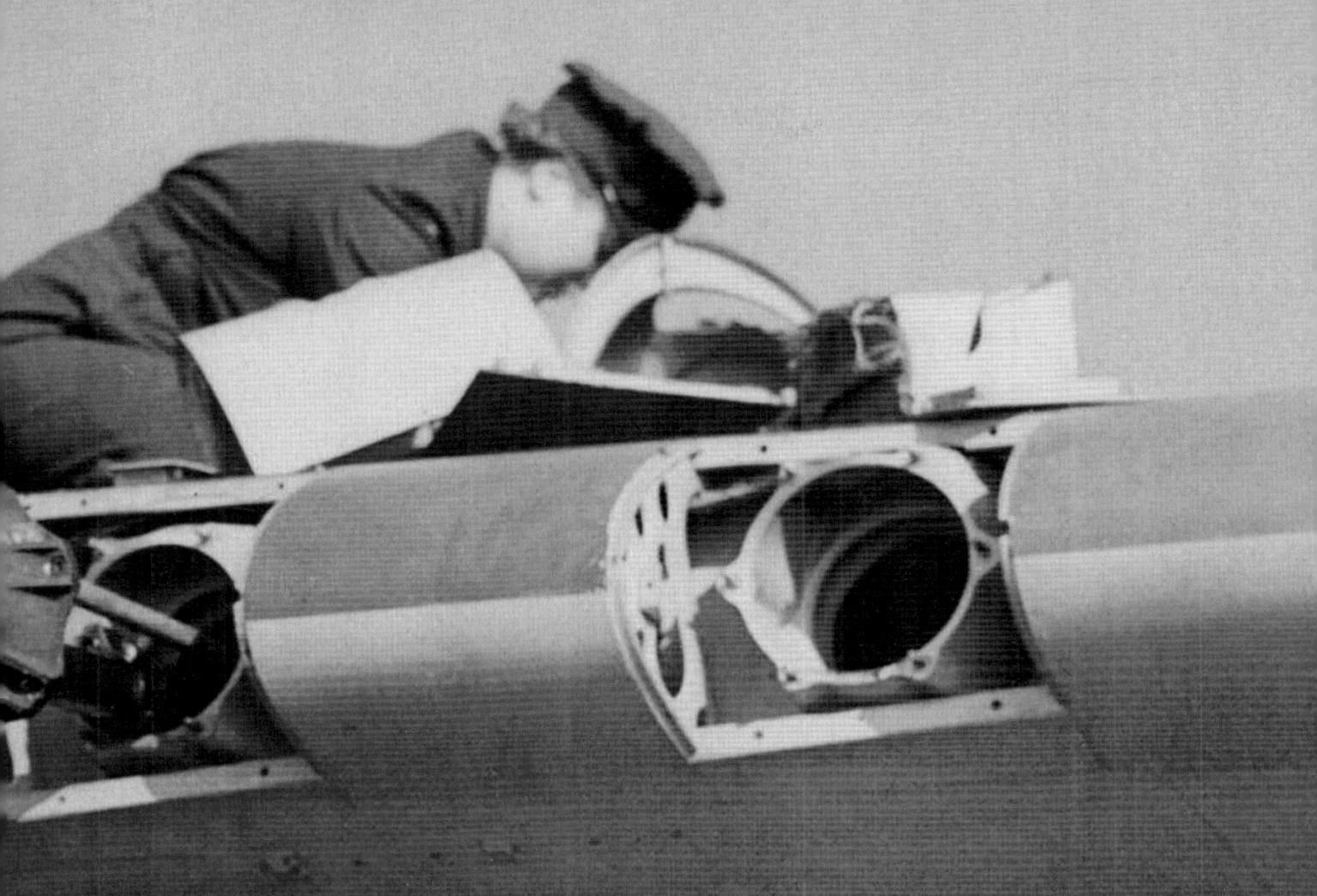

THE BIRTH OF AN ALMOST-LEGEND

HAWKER HURRICANE THE HISTORY OF A LEGEND

SYDNEY CAMM

'The main requirements of an aircraft designer (are) a knowledge of aerodynamics, some elementary maths, and an eye for beauty.'
– Sir Sydney Camm

'…one of the most consistently successful designers the aircraft industry has ever had.'
– The Times

'I can't imagine why his men put up with him.'
– Sir Thomas Sopwith

Sydney Camm, the designer of the Hawker Hurricane, was one of that quite incredible generation of British Victorians who were swept up in the sheer excitement generated by the invention of powered flight. The Wright Brothers had finally shown that it was possible and young men everywhere seized the opportunity to be part of something big and new and challenging. For an engineer just starting out on a career, there was no more appealing horizon.

Born the eldest of twelve children in 1893 in Windsor, Berkshire, Camm's father was a carpenter and, in 1908, Camm left school to follow him into the trade. He and his brothers made wooden model aeroplanes in their spare time and soon discovered that they could sell these for a profit to the pupils of nearby Eton School. The brothers would sneak into the school grounds by night for illicit assignations. The models would be hauled up by string into the dormitories and down would come payment in a receptacle by return. Camm promoted his models with slogans such as *'Will Really Fly'* and perhaps more modestly *'Will Rise from the Ground'*.

In 1912, Camm co-founded the Windsor Model Aeroplane Club. That same year, he and his fellow enthusiasts built their first glider capable of carrying a man aloft. They would later add an engine to their design.

Despite his obvious abilities in aviation design, it was his family background as a carpenter that first assured Camm a job in the industry in 1913. The Martinsyde aircraft company of Brooklands took him on as a shop floor carpenter. He was swiftly promoted to the drawing room as a designer and stayed with

✈ Sydney Camm at the Windsor Model Aeroplane club
✈ Hawker Cygnet biplane

THE BIRTH OF AN ALMOST-LEGEND

the company throughout the First World War working mostly on fighters.

Camm first joined H.G. Hawker Engineering Company Limited of Kingston-upon-Thames in 1923. He was initially employed as a Senior Draughtsman and, after his design for the two-seater Hawker Cygnet light biplane won several speed prizes, was promoted to Hawker's Chief Designer in 1925. Thus began a truly extraordinary career that saw him design no fewer than 52 different aircraft. It's said that, during one period in the 1930s, 84% of the aircraft being flown by the Royal Air Force were designed by Camm. By 1925, he was already thinking about a small monoplane fighter type, but conventional industry thinking and Camm's own conservative caution confined it squarely to his own head.

Around Hawker, Camm enjoyed a quite fearsome reputation as an eccentric tyrant. Hawker engineer Robert Lickey recalled:

'Camm had a one-track mind – his aircraft were right, and everybody had to work on them to get them right. If they did not, then there was hell. He was a very difficult man to work for, but you could not have a better aeronautical engineer to work under.'

Engineer John Fozard described Camm's legendary temper as *'bowel-turning'* and fellow sufferer Robin Balmer said that *'he was wonderful at motivating people, more by fear than anything else.'* Others on the team recalled him being pleasant enough when cheerful, but prone to deep depression too. Still others remarked on his awkward shyness, propensity to whistle and an almost pathological dislike of wind tunnel tests.

'I am only interested in designing fighters. There's no finesse in anything else.'
– Sydney Camm

Once Sydney Camm had a thought and had successfully distilled it into a phrase, he would feel compelled to share it with the engineers working under him in the Projects Officer. Again and Again. And then again for good measure. It drove most of them quite mad, as did the equally repetitive use of some favourite pieces of abuse such as *'my mother could have done that!'*

The more sympathetic of his men thought that Camm was, most cunningly, trying to forge a defensive esprit de corps in his design team. Less kindly fellows swore that 'the Old Man' was instead adversely afflicted by phases of the moon.

Of course, it was also entirely possible that Camm – a huge admirer of Evelyn Waugh – thought that he was being funny and entertaining and charmingly eccentric and missed the mark by several miles.

Over the decades, his team of engineers included men as accomplished as himself. These included Sir Frederick Page who created the English Electric Lightning, Leslie Appleton – the man behind the Fairey Delta 2 and Stuart Davies, Chief Designer on the Avro Vulcan project. All learned their trade from Camm.

Camm's own pre-Second World War war designs included the Hawker Tomtit, Demon, Audax, Osprey, Hornbill, Hart, Fury, Nimrod and, of course, the Hawker Hurricane. During the war years, he would be responsible for both the Hawker Typhoon and Tempest. The Royal Navy's carrier-based Sea Fury, first entering service in 1947, was another invention of Camm's. Entering the jet age, Camm created the Sea Hawk and another classic British fighter – the Hawker Hunter, before becoming an integral part of the Harrier VTOL project.

Camm was knighted on 2 June 1953 for his services to British Aviation. He only retired from his post as Chief Designer in 1965 and died scant months later. He was seventy-two years old and his career with Hawker had spanned 43 years.

'Oddly enough, he never liked flying, and on one occasion when he had to go to America he had to go by sea.' – Elizabeth Dickson, Sir Sydney Camm's grand-daughter

A GOOD PLANE IN A BAD YEAR

1934 was a particularly bad year to pitch the idea of a new monoplane fighter aircraft to the British Air Ministry.

Despite the Nazis' rising to power in Germany the previous year, there was a particularly strong element within parliament committed to the cause of peace and even progressive disarmament. They wanted a better world, not weapons of war. They wanted the

L-R: Hawker Tomtit; Sir Frederick Page; Hawker Nimrod
A British Air Ministry meeting
Sopwith Camel

THE BIRTH OF AN ALMOST-LEGEND

new League of Nations help make all nations brothers. So what if Germany had just left the institution with almost indecent haste? No-one wanted to see another World War after the horrors of 1914-1918, and so there was an almost religious belief – a faith – in the idea that it couldn't happen again. The road to Hell was thus paved with good intentions – and trod by optimists. Even amongst those of cynical mind, there was the pressing reality of a Britain in the 1930s which simply could not afford to rearm. Money was scarce. Why build fighter planes when so many poor lived in abject squalor? This sentiment was not shared by Nazi Germany, where the head of the Luftwaffe famously said, *'guns before butter.'* In 1933-4 they were busying themselves designing superior fighters and bombers and introducing national conscription. Back home, Labour and the Liberals formed a pact to stop the RAF being expanded.

Even amongst the military establishment, who were not known for turning down new weaponry when it became available, there was considerable confusion

about what they really needed. At the Air Ministry, the older men tended to still favour the biplane. Monoplanes were a passing fad. They remembered the good old days of the Royal Flying Corps and the Sopwith Camel. Biplanes were more reliable, more manoeuvrable and simpler to control.

HAWKER HURRICANE THE HISTORY OF A LEGEND

Younger members of the Ministry believed on the other hand that the future lay with the monoplane. They pointed to Schneider Trophy-winning machines from Supermarine. They believed in speed – a true young man's passion in the 1930s – and the Schneider racers had certainly delivered that, exceeding 400mph in one case.

The monoplane vs biplane debate was one bone of contention within the Air Ministry, but so was the fracture between the supporters of the bomber and those of the fighter. *'The bomber will always get through,'* proclaimed Stanley Baldwin in 1932 – a view held by those in power since the very earliest days of 'Boom' Trenchard's Royal Air Force. Fighters were seen as a waste of time. The many-gunned bomber would simply bully its way to the target, swatting fighters aside like so many flies and deliver a knock-out blow to the enemy on the very first day of conflict. That's why, in the early 1930s Fighter Command consisted of a meagre thirteen squadrons. The Air Ministry ran on orthodoxy just as much as it ran on cups of tea and due deference for your social superiors.

No money. No vision. No unanimity and no enthusiasm. So, 1934 was a particularly bad year to pitch the idea of a new monoplane fighter aircraft to the Air Ministry – but Sydney Camm decided to try it anyway.

On 6 January 1934, Camm submitted his proposal for a new 'High speed Single Seater Fighter'. It packed four machine guns (two on the wings and two in the fuselage) and promised a lot – including speeds of 264 mph. The Ministry were aware of Camm's interest in pitching a new fighter. They'd been in discussions with him over the past year about the vague possibility of supporting a monoplane version of the Hawker Fury, called variously the *'Hawker Monoplane'* or the *'Fury Monoplane.'* This would have had the Fury fuselage at its heart and the most significant change would have seen the top wing removed. It would

✢ Hawker Fury
✈ Sydney Camm in his office going through designs
✢ Rolls-Royce Merlin engine

THE BIRTH OF AN ALMOST-LEGEND

feature the basic Hawker fabricated steel tubular structure with fabric covering in most areas. The cockpit would be enclosed because of the greater speed anticipated from the design.

The Air Ministry duly responded to Camm's proposal on 9 March 1934:

'It is regretted that, at the present time, the Department is unable to give active encouragement to the scheme proposed.'

Five months later they rejected the Spitfire.

Camm's response was not to give up but to try harder. He started to abandon Hawker's tried but trusted biplane thinking and to forge ahead with new technology, such as a retractable undercarriage. The power plant for the original design, the Rolls-Royce Goshawk engine, was replaced because of its poor performance in tests. The new engine that Camm now chose was the Rolls-Royce Merlin.

Having ploughed ahead on his own, Camm went back to the Ministry on 3 September 1934 with detailed plans

for an 'Interceptor Monoplane'. He said he recognised that the Ministry had rejected his last effort, but his new design incorporated many improvements. If the RAF wanted it, Camm was confident that they could have it by mid-1935. For some reason this time the Ministry looked upon Camm with more favour. They ordered a prototype – provided Hawker would accept £8,000 for it. At the same time, parliament was becoming far more nervous of the looming threat from Nazi Germany and proposed greatly boosting the size of the RAF to match.

HAWKER HURRICANE THE HISTORY OF A LEGEND

✈ ✈ Hawker Hurricane prototype K5083

In November 1934, the Ministry issued a new specification – F5/34 for a monoplane with eight Browning machine guns. Tacticians had decided that the best way for a fighter to bring down a bomber was heavy firepower. The Hawker Prototype was intended to have only two machine guns (later rejigged to four), and by now the design was too advanced to meet the new specification. Things progressed in parallel. Hawker unveiled a non-flying mock up of the 'Hawker High Speed Monoplane' to officials in January 1935. The Ministry liked what it saw and gave the go-ahead for a flying prototype and even drew up an exclusive specification – F36/34 for Hawker's new fighter to meet. F10/35 likewise brought Supermarine's Spitfire back from the dead.

Squadron Leader Ralph Sorley, a leading advocate of the eight gun fighter, inspected the progress of both the Spitfire and Hurricane teams in 1935 and reported that *'both aircraft look to be excellent.'* He suggested that both should go into production immediately and prototypes be damned. This was too much for the Ministry to consider – but *suddenly the Hurricane was being spoken of in the same breath as the Spitfire*. Suddenly, it was important. By 13 May 1935, the previously obscure Hawker fighter found itself being discussed in parliament, as frightened MPs looked for any way expand the RAF for war in double-quick time.

Hawker worked with great vigour through the summer and autumn of 1935 to complete their prototype which, according to one member of staff, was shaping up to be an *'aggressive and menacing-looking machine'*. Politicians frantically chased its progress, demanding to know just how soon it could fly. The answer would be 6 November 1935.

THE BIRTH OF AN ALMOST-LEGEND

'PIECE OF CAKE!'

'When the design of the Hurricane had gone beyond the point of no return, I suddenly had a foreboding that it would be no good.'
– Sydney Camm

The Hurricane was still nameless when it was rolled out at Brooklands for its first test flight on the chilly morning of 6 November 1935 – but looked splendid in its all-silver finish. Hawker's Chief Test Pilot, Paul 'George' Bulman MC, took what was then just the unglamorously designated K5083 up for half an hour and put the aircraft through its initial places. *'Piece of cake!'* he announced to Camm after completing the flight. This wasn't strictly true, since Bulman had found both taking off and landing somewhat sticky, but overall the aircraft had proved safe and capable.

Seven more test flights followed. During one, the cockpit canopy came away, bouncing off the fuselage before plunging to earth. 'George' Bulman's lucky trilby hat, which he always tried to wear on test flights, was sucked straight off his bald head and irretrievably lost. There were also problems experienced with the retractable undercarriage pump and more importantly, with the new Merlin engine. Almost every new test flight saw the installation of a new Merlin.

Meanwhile, an increasingly nervous Air Ministry nagged incessantly for the F36/34 to be turned over for official RAF trials at Martlesham Heath in Suffolk. Appetite had replaced apathy and the new fighter simply couldn't be pressed into service quickly enough. Hawker bridled at the pressure being applied to them, wanting to do more thorough appraisals of their own. The Ministry didn't care. Within scant weeks of the first test flight, they were already talking to Hawker about placing their first order.

The press fanned public excitement. The Daily Express reported that the new fighter would be faster than a seagull, but then blotted their copybook by referring to the new aircraft as *'The Merlin.'* Flight Magazine celebrated the advent of *'300 mph machines'*.

F36/34 was finally delivered to Martlesham Heath for independent RAF assessment on 7 February 1936. Along with it came a promise from Hawker that if

HAWKER HURRICANE THE HISTORY OF A LEGEND

needs be, the first aircraft off the production line could be available by June 1937 with 40 planes a month thereafter. The usually-conservative Ministry took what they called 'a gamble' and proposed to Hawker that same month that their initial order might be for a quite incredible 600 fighters, Now the delivery date of June 1937 became a bone of contention. Dissenting voices asked why it couldn't be even sooner. It was suggested that the task of building the new F36/34 could simply be taken away from Hawker and given to a bigger, better aviation company. Hawker came back with a counter proposal, saying that the first aircraft might be feasible by April 1937 instead of June, with the whole order of 600 fighters delivered by June 1938. By March 1936, the order for 600 planes became official, despite the fact that the RAF were not finished with their appraisals. That same month, the Spitfire had its first test flight.

✝ Spitfire prototype K5054

It was fortunate for all concerned then, that when the trials at Martlesham finally ceased in April 1936, the F36/34 passed with flying colours. The report concluded;

'The aircraft is simple and easy to fly and has no apparent vices. All controls operate satisfactorily...'

At the beginning of June, the same month that the Ministry produced an order for 419 Spitfires, Hawker received an official government contract for an initial 600 planes. That same month, the British Public got to see the new fighter for the first time when it performed demonstrations of speed and aerobatics. There was immediate clamour to find out what this glamorous new fighter plane would be called. Hawker duly proposed that the F36/34 should henceforth be officially known as the Hurricane. The

Ministry approved and King Edward VIII presided over an official naming ceremony at Martlesham the following month.

Because of the sheer speed with which the whole project had to be completed, one wag suggested that the plane should be nicknamed the 'Hurry'. The name stuck.

READY, STEADY…

'If we'd had more time it would have been the greatest aircraft of all time.'
– Sydney Camm

Hawker now had the difficult task of living up to their promises and delivering on time what had been the largest-ever aviation order in peacetime. Indeed, they had made matters worse for themselves. In secret they had upped the initial production run from 600 to 1,000, confident that good export sales would soon account for the difference.

Hawker began to expand. They had already acquired Gloster in 1934, but their facility was still preoccupied with building the Gloster Gladiator. Thus it was in July 1936 that Hawker merged with the Armstrong Siddeley Group to become Hawker-Siddeley. At all times, the Air Ministry kept up the pressure, regularly pointing out perceived inefficiencies on Hawker's part. There were even fears expressed that Camm would eventually 'crack up' under the strain and need replacing. The Air Ministry's tetchiness was further exacerbated by Hawker putting the availability of the first Hurricane back to first July and then September 1937.

At the heart of it, Hawker's plant at Kingston was not suited to mass production. There was just room to work on five Hurricane fuselages and wings at any time, and these then had to be transported to Brooklands for final assembly and testing. Camm

✈ King Edward VIII
✈ Gloster Gladiators

was always a fan of speed in all things including work, but here he was up against real physical obstacles. Luckily for Hawker, the Air Ministry understood the problem and advanced them a loan of £30,000 to buy a new site in Langley, Berkshire. It was an extraordinary piece of foresight on the part of the Ministry. Although the plant wouldn't be ready until 1939, it would eventually turn out some 7,000 Hurricanes over the course of production. The new site dwarfed both Kingston and Brooklands. *'It was a different world'*, one Hawker employee later recalled. And it made a real difference. The first Hawker Hurricane came off the production line in October 1937. It was given the serial number L1547. Even more serious delays with the Spitfire saw the first production model held up until mid-1938.

PRODUCTION

Even as the Hurricane went into production, changes were still being made during the run. Perhaps most important of all, Hawker stopped specifying the Merlin I engine and switched to the Merlin II. Because it was a slightly different shape and size, further alterations had to be made to the nose of the aircraft. Everything from the engine mounting and air-intake to the

✈ Production of Merlin engines at the Rolls-Royce factory

✈ Hurricane outline drawing

propeller and the glycol tank had to be redesigned for conversion. The change added four months to the production schedule.

Change seemed to be a constant during the first and second production runs. The old Watts two-bladed fixed-pitch wooden propeller used – a real throwback to the days of the biplane – was scrapped and in its place came first a de Havilland variable-pitch three bladed propeller and then a constant-speed Rotol propeller. This helped boost both performance at low speed and to climb rate, and to make take-off easier and safer in a shorter distance. Fitted with the new propeller, the Hurricane could take off safely in just 250 yards compared with 410 yards with the old propeller – an important consideration given the size of many RAF aerodromes.

In came an all-metal, stressed-skin wing instead of fabric which allowed for far faster diving speeds. The rudder was increased in size. In came better exhausts and improved radios. Lowering the undercarriage was now powered by the engine rather than the brute strength of the pilot. More efficient 100 octane fuel replaced the old 87 octane, further increasing speed and climbing rate. The Merlin II gave way to the Merlin III.

The cockpit had been redesigned since the prototype and at Dowding's insistence, the canopy windshield was made bulletproof. Armour plating was added in critical areas to the front of the cockpit to offer vital protection if the pilot was chasing a bomber and facing its rear guns. Because it was not considered that the Hurricane would have to contend with a

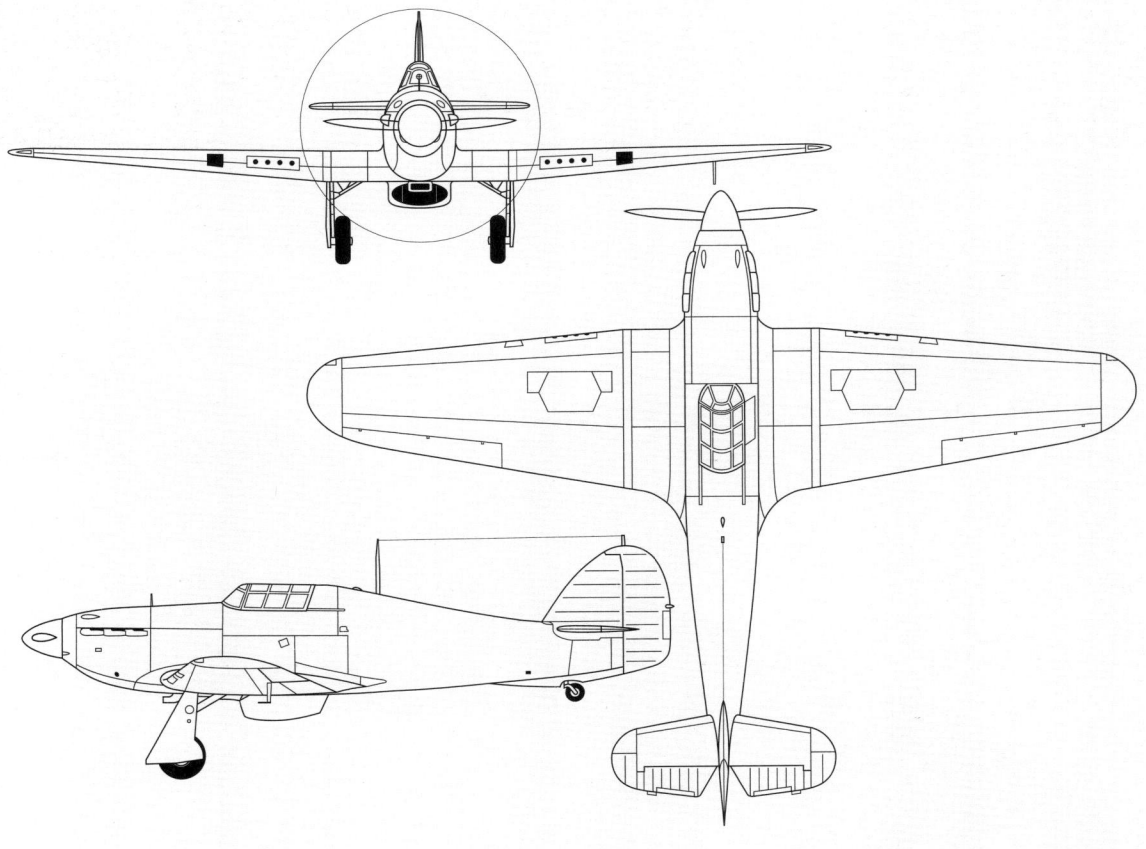

fighter chasing it from the rear, armour plating was still not fitted to the back of the cockpit.

(It was No 1 Squadron fighting out in France who realised this was a serious vulnerability. They fitted their own makeshift armour plate, salvaged from a crashed Fairey Battle, into the critical spot. A pilot, Flying Officer Brown, was then selected to fly the Hurricane back to Farnborough with the suggestion that such an arrangement might be applied to all future Hurricanes. The RAF duly took notice and started to convert existing Hurricanes with this extra protection as quickly as possible.)

'SECOND BEST'

Just when it seemed that everything was going well with the Hurricane project, the Air Ministry suddenly had second thoughts about the aircraft they had previously (and impatiently) championed. The Spitfire prototype was flying now and clocking up speeds in excess of 35 mph more than the Hurricane. Suddenly the Spitfire was the shining hope of the Empire and the Hurricane a dumpy second best. The Ministry now regretted ordering 600 Hurricanes and just 300 Spitfires. They had, it was feared, bought a pig in a poke; A slowcoach in an era of speed.

It was the C/O of Fighter Command, Hugh Dowding, who went into battle for Hawkers, explaining that the Spitfire took far longer to build and they could have more Hurricanes ready for battle soonest. Although he didn't know it, Dowding was more prescient than he thought. The Spitfire project would be plagued by all sorts of problems in the next few years, slowing down production by even more than the most pessimistic onlookers were counting on.

INTO RAF SERVICE

The very first Hurricane was handed over to No.111 (Fighter) Squadron based at Northolt, Middlesex in

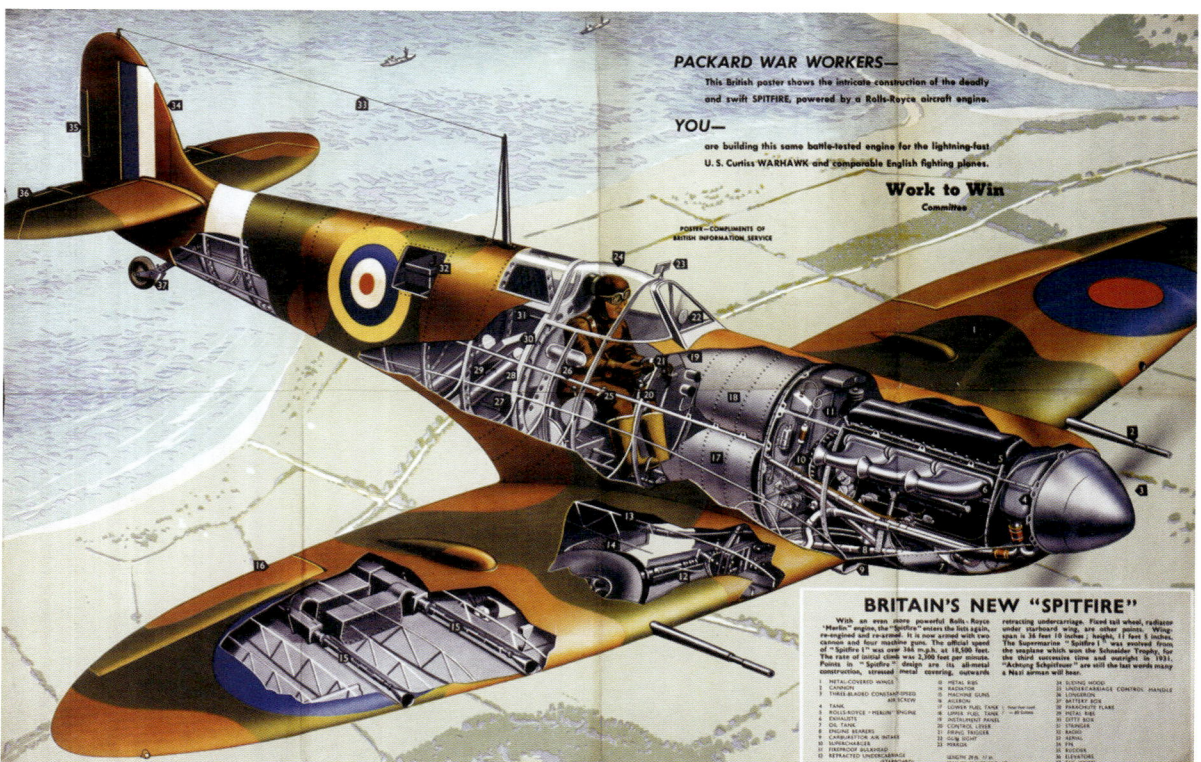

THE BIRTH OF AN ALMOST-LEGEND

December 1937. Four more arrived soon after and the squadron were able to form their first Hurricane flight, converting from their ageing Gloster Gauntlet biplanes. Twelve more joined the squadron by February 1938.

There was some trepidation amongst the pilots about conversion to the Hurricane, particularly at coping with the increased speeds their new aircraft could offer. They were well aware that they were guinea pigs and there was at least one fatal accident. Any aerobatics were banned. Rumours spread that only 'super-humans' and senior officers could fly a Hurricane.

To try and settle his mens' nerves, on 10 February 1938 111 Squadron Leader John Gillan flew his Hurricane from Edinburgh back to Northolt with no discernible problems, achieving an average airspeed of 408.75 mph. At one point close to home he put the aircraft into a dive and notched up 450 mph – a speed so unexpected that he almost overshot his home base. There were jokes in the mess about Gillan enjoying a very favourable 80 mph tailwind to speed him on and the Squadron Leader was given the new nickname of *'Downwind Gillan.'* Whatever the truth, the Hurricane had just accidentally shattered the landplane world speed record, an achievement previously held by its arch-rival the Bf-109 and was now the fastest warplane on earth.

The daily press covered the story and so the wider world thought it a fantastic achievement. The Hurricane now enjoyed the excitement and acclaim that would later belong exclusively to the Spitfire. The Times lavished praise:

'(This) is a record of which no previous fighter has been capable, and marks the Hurricane as outstanding in its class in respect of duration as well as speed.'

✤ The Spitfire was proving to be popular
✤ Northolt air base
✤ Hugh Dowding, who stood up for the Hurricane

HAWKER HURRICANE THE HISTORY OF A LEGEND

111 Squadron enjoyed an enhanced reputation as the cream of the RAF for a while. Flight Magazine called them *'the most formidable fighting squadron in the world.'* The great, the good (and the ordinary) flocked to catch a glimpse of the new Hurricanes of Northolt. 111 even enjoyed a visit from his Majesty King George VI and invitations to flying displays both domestic and international.

In March 1938, No 3 Squadron at Kenley received their first Hurricanes to replace their Gloster Gladiators, but following another fatal crash they decided that their runway was too small to adequately allow a Hurricane to take off and switched back to Gladiators once again, They would not adopt the Hurricane again until May 1939, by which time they had moved aerodromes. The following month, 56 Squadron did successfully convert from Gladiators to Hurricanes and 71 and 87 Squadrons converted in July.

Despite the nerves and mishaps, the Air Ministry were pleased enough by the introduction of the Hurricane to order 1,000 more from a second production run on 1 November 1938. War was less than a year away and those in power knew it, even if the public chose to believe Chamberlain's *'Peace in our Time'* speech.

Things were still not going well with Spitfire production and it seemed like it was the Hurricane or nothing. By Spring 1938, not one single Spitfire had been built.

By the end of 1938, the RAF could field ten Hurricane squadrons. By the time war was declared on 3 September 1939, eighteen squadrons had received Hurricanes and three more were in the process of converting. The RAF could now put 497 Hurricane fighters into the air.

✈ TOP: King George VI L-R: Gloster Gauntlet; Women working hard assembling Hurricanes; Production of the German Messerschmitt Bf-109

✈ A Nazi Germany ceremonial parade INSET: British Prime Minister Neville Chamberlain (right) and the leader of the Nazis, Adolf Hitler (left)

THE BIRTH OF AN ALMOST-LEGEND

REAL WAR/ PHONEY WAR

HAWKER HURRICANE THE HISTORY OF A LEGEND

DEPLOYMENT TO FRANCE

In response to the outbreak of war, ten army divisions, both regular and territorial, were dispatched to France over time nominally under General John Standish Surtees Prendergast Vereker, 6th Viscount Gort (but actually controlled by the French High Command.)

Initially four Hurricane squadrons were sent to France, much to the annoyance of the French authorities who were calling for a commitment of at least ten squadrons in the first instance. 85 and 87 Squadrons (60 Wing) were deployed to support the British Army contingent of the British Expeditionary Force, whilst 1 and 73 Squadrons (67 Wing) were assigned to provide fighter cover for the British bombers given over to the Advanced Air Striking Force. 607 and 615 Squadrons were to join the BEF contingent in November of 1939. More would follow.

The Hurricane squadrons were greeted by the French in most cases with a mixture of indifference and perhaps even contempt. Transport arrangements for the pilots and ground crew were shambolic and billeting facilities ranged from the filthy to the non-existent. Communications were dire. When the C/O of 73 Squadron took his men up for a welcoming salute, they were accidentally fired upon by the French coastal batteries. Fortunately, their aim was as poor as their aircraft identification skills.

None of this augured well, but none of this also came as a surprise to the Head of Fighter Command, Hugh

REAL WAR/PHONEY WAR

Dowding. He had made a courtesy peacetime visit to the French Air Force at the beginning of 1939 and found them in absolute chaos, although the meal he received was first rate. He was to return home in a foul mood. He feared – with good reason – that any war fought in France might leave the RAF *'bled white.'*

Despite the chaos and disinterest in France, the Hurricanes quickly scored their first ever kill of the war, when on 30 October 1939 Pilot Officer P 'Boy' Mould of 1 Squadron brought down in flames a Dornier Do-17 on a reconnaissance flight. He had spotted the bomber from the ground, taken off without permission and shot it down at 18,000 feet on his own initiative.

Another Do-17 was claimed on 8 November by Flying Officer Edgar 'Cobber' Kain of 73 Squadron, who chased it down from 27,000 to almost zero feet during a dogged pursuit that lasted for a full quarter of an hour. Kain – a New Zealander – would claim four more kills in his Hurricane by March 1940

✈ Hurricane being prepared for take off in France

and distinguished himself as the first RAF 'Ace' of the Second World War. Three of his 'kills' were enemy fighters.

The 'Phoney War' in France echoed in many ways the experiences of the Hurricane pilots back home. Enemy activity was sporadic at best, and usually involved lone or else small numbers of German bombers, sometimes acting in a reconnaissance role. The two squadrons posted close to the Maginot Line – 1 and 73 – were to see the bulk of the fighting at this time, while further north 85 and 87 squadrons were required to do little more than huddle round their stoves and keep warm. That's not to say they didn't get their own 'Kills'. 87 Squadron managed its first kill – an He-111 on 2 November, while 85 got its first 'kill' – also an He-111 – on 21 November.

The winter of 1939/40 was foul, with snow piling up so high on the French aerodromes that pilots could become stuck in their freezing cold huts. This severely limited the flying on both sides and when a Hurricane could make it into the air, its total lack of any form of cockpit heating just added to the misery.

On 22 December 1939, the Hurricane tangled with its anticipated nemesis, the Bf-109, for the first time. Three Hurricanes from 73 Squadron were ambushed by four German fighters. In the resulting dogfight, two of the Hurricanes were shot down for the loss of one German fighter. Both RAF pilots died. However, the analysis of the duel showed that the Hurricanes had acquitted themselves well, given that they had been taken by surprise. The Bf-109s would find

✈ German Dornier Do-17s; Heinkel He-111

✈ 'Cobber' Kain stood in front of his Hurricane Mk I 'Paddy III'

REAL WAR/PHONEY WAR

them no pushover. It was rivals No 1 Squadron who got the glory on 9 March, shooting down a Bf-109 in the morning and a twin-engine Bf-110 'Zerstorer' (Destroyer) in the afternoon.

March 1940 saw the weather improving and air activity increasing. On 2 March, No 1 Squadron succeeded in bringing down a Do-17, but not before one of its gunners had raked a Hurricane with return fire. The Hurricane crashed on landing, killing its pilot. The next day, 1 Squadron added an He-111 to their tally. That same day, 73 Squadron were on patrol when they spotted a formation of seven He-111s. 'Cobber' Kain was amongst them. Without consulting anyone, he peeled away to chase the bombers and raced after them, leaving his fellow pilots trailing behind.

Such was his excitement that he didn't realise he was now flying over Germany itself – and that a swarm of Bf-109 fighters were sitting on his tail and lining him up in their sights. His wingman shouted a desperate warning over the radio and Kain took abrupt evasive action. As the Bf-109s swept over and past him, he managed to get one in his sights and brought it down. Still blazing away, he damaged two more Bf-109s. Seconds later, his own aircraft was taking multiple hits almost everywhere. Its engine failed. Kain, trapped in a cockpit filling with smoke, aimed his nose at where he hoped his home base might be and started to glide his Hurricane down. On the way, he spotted a French airfield and settled on landing there, jumping out of his cockpit just as his Hurricane burst into flames.

HAWKER HURRICANE THE HISTORY OF A LEGEND

By the end of March 1940, Luftwaffe fighter activity was noticeably still increasing in the skies over France. They came with greater frequency and in greater strength. On 26 March, 'Cobber' Kain's luck finally ran out – but only after claiming two more Bf-109s. His Hurricane was then badly shot up and caught fire, but Kain managed to parachute to safety and won his second DFC in the process. His fellow 73 Squadron pilot, Flying Officer Orton, also shot down two Bf-109s in the same fight. On 29 March. No 1 Squadron again tangled with Bf-109s and 110s. The squadron shot down one of each but also lost a Hurricane.

THE WAR AT HOME

'There was lots of sitting around. It was rather tedious.'
– Hurricane pilot Charlton Haw

On the Home Front, in the early months of the war, relatively little happened of any real significance in the air. Strikes when they came were mostly against naval facilities and shipping. In Britain, squadrons filled much of their time with practice and exercises or else 'convoy cover' (which one pilot described as *'terribly dull'* and another as *'(just') going round and round.'* Flying was often curtailed by freezing temperatures, snowstorms or thick fog in the winter months. There was no small degree of indecision, confusion and tragedy.

Early on the morning of 6 September 1939, Hurricanes from 56 Squadron were scrambled to hunt down possible high altitude German raiders. Due to some confusion, 151 Squadron were also sent to attack the same target. Radar operators saw the two friendly Hurricane squadrons on their screens and reported them as a sizeable enemy force. Anti-aircraft guns began to open up on the Hurricanes. 11 Group's headquarters at Uxbridge then scrambled three squadrons of Spitfires to intercept the Hurricanes. 74 Squadron reached the Hurricanes first – and shot two of them down. Pilot Officer Frank Rose survived a crash-landing, but his fellow Pilot Officer, Montague Hulton-Harrup, was not so fortunate. He was killed instantly when fire from one of the Spitfires shattered his cockpit and his Hurricane plunged into the ground. He was the first RAF pilot to die in the Second World War. The tragic 'Friendly Fire' incident became known as 'The Battle of Barking Creek.'

No 43 Squadron's first mission was to shoot down a barrage balloon from London that had broken loose.

✈ L-R: Directing war operations; The Home Front in Britain; Ground crew and pilot waiting to see some action in the air

✈ Hurricanes heading out on a sortie over England

REAL WAR/PHONEY WAR

It apparently took Flying Officer John 'Killy' Kilmartin 1200 rounds from the Hurricane's guns. This was far from a rare occurrence and squadrons soon had a competition going as to who could shoot down the most rogue balloons. No 111 Squadron claimed that they were the RAF champions, having accounted for seven balloons in just a single day.

When the Germans did come, it was usually to spy on – or attack – naval targets in North-east England or Scotland. On 21 October, Hurricanes from 46 Squadron intercepted a flight of He-115 seaplanes, shooting down three and forcing a fourth to crash land on the North Sea. The first German bomber casualty was claimed by 79 Squadron, who brought it down over the Channel. No 111 were posted to Northumbria in November 1939 and soon brought down a Heinkel He-111 over the North Sea. Similar encounters followed. 43 Squadron got their first (non-balloon) kill – an He-111 – on 3 February 1940. It crashed in snowy fields outside Whitby in Yorkshire with three of its four crew dead in the wreckage.

On 8 April 1940, Hurricanes from No 43 Squadron, charged with providing air cover for the naval base at Scapa Flow intercepted two waves of incoming German bombers. In the melee that followed, they succeeded in shooting down three He-111s for no loss. A fourth crash-landed at the Hurricane's home airfield, with half the crew on board dead. In the dark, the German survivors assumed they had just landed on water, threw out an inflatable dinghy and then dived out of the aircraft into what they anticipated would be water. Bemused onlookers scratched their heads and laughed.

On 10 April, a mass air battle occurred between Hurricanes and German raiders over the Orkneys. The Hurricanes claimed seven 'Kills'.

BY CARRIER TO NORWAY

On 9 April 1940, German forces invaded Norway to secure the nation's vital iron ore resources. Britain was caught largely unprepared and the troops it despatched to try and counter the German forces were badly beaten in short order. At the start of May, the survivors were evacuated in chaotic conditions.

One of the main reasons for the defeat was understood to be lack of Allied air power to support the British troops. Of all the aircraft the RAF had under its command, only the fragile Blenheim had the range to take it to Norway and even then it had scant minutes to strike before having to return home. The initial solution was seen to be conveying Gloster Gladiator biplanes across the North Sea by aircraft carrier. Eighteen of the outdated aircraft were taken across by the aircraft carrier *Glorious* and then flown off to a makeshift airstrip and base on a frozen fjord near Aandalsnes. Within a week, all eighteen had either been destroyed or else disabled by the cold. Something better was desperately needed. Special Hurricane floatplanes were considered but would take too long.

It was then decided to ship out the Hurricanes of 46 Squadron under the command of Squadron Leader Kenneth 'Bing' Cross, again on board the carrier *Glorious* and then fly them off to a new land base up in the far north near Narvik. No one knew if Hurricanes could operate from a carrier, but matters were so desperate that there was no choice except to press ahead and find out the hard way.

As no-one had ever attempted to land a Hurricane on a carrier before, the aircraft were loaded onto barges and then laboriously hauled up onto the *Glorious* by the ship's cranes as she waited on Clydeside. They were then transported across the North Sea to await deployment.

On 26 May, the signal came that they were good to go. Leading by example, 46 Squadron's commander

✈ L-R: A Bristol Blenheim over the North Sea; HMS *Glorious*
✈ Devastation in the Norwegian port of Narvik

made the first attempt to take off from the *Glorious*. If he didn't survive, then perhaps there might be second thoughts. Cross started up his Hurricane, hammered down the deck and to his delight (and perhaps to his surprise), his Hurricane left the ship with power and room to spare. The real problem came when the Hurricanes tried to land on their new improvised airstrip at Bardufoss. It proved so rough that several Hurricanes were damaged in the attempt, including Cross's. By all accounts, he laid into the officer in charge there with such ferocity that he made him cry. The Hurricane pilots were horrified by the state of their new airstrip, where they had to sleep six to a tent and bathe in '*a puddle of snow*'.

They were in action almost immediately helping army and navy forces to seize the port of Narvik from the German invaders. By 2 June, 46 Squadron had accounted for eleven German warplanes while the Germans occupying the port had been driven back all the way to the Swedish border. And then the British commanders gave up. With the German Blitzkrieg raging through France and the Low Countries, they decided that Britain simply couldn't fight a war on two fronts. On the night of 3-4 June, British forces began to evacuate, with the Hurricanes of 46 Squadron giving them air cover.

Ten airworthy Hurricanes now remained. Cross was given the choice of destroying them or else try to fly them out and perform an untested landing on the *Glorious*. Ground crews would be evacuated separately. Cross decided that the RAF couldn't afford to lose ten Hurricanes so he elected to fly them out. It was a truly desperate thing to do. Taking off from a carrier was one thing, but landing on one might prove to be quite different. The weather could be atrocious, the ship could be pitching and rolling in harsh seas and the Boffins said that a safe landing couldn't be done. Cross decided nevertheless that his squadron would take off in two formations on the 8/9 June. The first would comprise three Hurricanes. If they made it

and landed safely the others would be told to follow. Incredibly both waves of Hurricanes made it with no serious mishaps, against all the odds.

With 46 Squadron safely on board, the *Glorious* turned and headed back to Scapa Flow together with her two destroyer escorts, the *Ardent* and the *Acasta*. Having been awake and flying in the most hazardous fashion for over 24 hours, the exhausted Hurricane pilots took the opportunity to finally relax and catch up on their sleep below decks. They thought they were safe and going home...

Unfortunately, that same fateful day, 9 June, the funnel smoke from the *Glorious* and her escorts caught the attention of a German raiding party comprising the battleships *Scharnhorst* and *Gneisenau,* plus a heavy cruiser and four destroyers, who immediately turned in pursuit at a quarter to four that afternoon. Fifteen minutes later the Royal Navy vessels spotted their pursuers and *Glorious* sounded Action Stations. The destroyer *Ardent* peeled off to investigate. As the opposing ships closed to a distance of 15 miles, the *Scharnhorst* opened up on the *Ardent* with her massive 11 inch guns. The *Ardent* got off a return shot, hitting the *Scharnhorst* and then laid down torpedoes and a smoke screen in order to escape – but to no avail. The *Scharnhorst* tore into her and she sank at 5.25pm.

The German battleship then turned her attention to the carrier. The *Scharnhorst's* big gun batteries missed the Navy carrier with her first two salvos, but on her third attempt she hit the ship on her forward flight deck, causing a catastrophic fire. A second shell hit the bridge of the *Glorious* killing or wounding almost all of the personnel. *Gneisenau's* guns joined in. A later salvo struck the carrier's engine room. She began to slow down and develop steering problems, then started to list. She was finished. Aboard the stricken carrier, the Hurricane pilots fought alongside the navy crew below decks to escape a wildly listing fiery death trap. Estimates say that around 900 men managed to get off the *Glorious* and into the water. Most then perished from exposure in the icy seas.

In one final grand pyrrhic gesture, the surviving destroyer *Acasta* burst out of her own protective smokescreen and let loose with her torpedoes, one of which hit the *Scharnhorst* towards the stern causing serious damage. The *Acasta's* guns scored a second hit on the *Scharnhorst,* but the small destroyer had

virtually no chance and was shot to pieces by the German ships. She sank at 6.20pm. History records just one survivor from each of the two Royal Navy destroyers and just 40 men from the *Glorious*' crew of 1,247.

Cross and one of his other pilots had scrambled through the dying ship and had managed to make it onto a life raft. No other pilots from 46 Squadron escaped. They drifted for three days, with other survivors dying all around them from their wounds or the cold. Both men eventually made it home. Cross would go on to command Hurricanes with distinction in the Middle East from 1941 to 1943, and received the Distinguished Flying Cross as well as Norway's highest honour for gallantry before ending his career as an Air Chief Marshal.

THE BATTLE FOR FRANCE

By the start of May 1940, Hurricanes from 1 Squadron had claimed 26 enemy aircraft while 73 Squadron had bagged 30. Things were hotting up.

Despite the increased air activity, no-one was prepared for the sheer force of the attack when Hitler launched his 'Blitzkrieg' on 10 May 1940. Suddenly, German bomber fleets seemed to be everywhere with their massed fighter escorts, providing close ground support for the panzer forces now racing through the Low Countries and out of the forests of the Ardennes.

RAF Hurricane fighters were in action from dawn on the first day, flying mainly to defend their own and nearby airfields from enemy bomber formations. They were joined by the Hurricanes of 503 Squadron who flew out from England and straight into a formation of 40 marauding Heinkel 111s. On the first day of combat, 49 German aircraft were claimed as destroyed or damaged by the Hurricanes – but it proved to be an exaggeration. They had fought well but were simply overwhelmed by the sheer number of German aircraft in the skies. 15 Hurricanes were lost on the first day – and that was not an exaggerated figure. Without reinforcements, they faced being totally overwhelmed. Their airfields were raided time and time again and they were continually pitched against vastly superior numbers. Dennis David, C/O of 87 Squadron vividly recalled his six Hurricanes fighting to save a hospital which was being dived-bombed by 40 Ju-87 Stukas. They succeeded in destroying 14

✈ L-R: German Panzer tanks were hotting things up in France; 87 Squadron pilots scramble to their Hurricanes; Junkers Ju-87 Stuka

✈ HMS *Ardent*

of the dive bombers, only to have another 150 Stukas suddenly appear to continue the raid.

As the 'Balloon went up', 3, 79 and 504 Hurricane Squadrons were immediately despatched to further support the BEF while 501 Squadron Hurricanes joined the Advanced Air Striking Force. 32 more Hurricanes followed over within four days.

Everyone knew that these numbers were far too few when pitched against the German onslaught. By the third day of the Blitzkrieg, the RAF had lost 96 Hurricanes. They were to lose a further 27 Hurricanes on 14 May – only the fifth day of concerted air engagements. The losses were clearly unsustainable.

Adolf Galland, Germany's most famous fighter pilot, did not hold the Hurricanes he fought in high regard

'*We outstripped them in speed, in rate of climb, in armament,*' he wrote in his biography. He neglected to point out that he Hurricane was a steady and formidable gun platform, could out-turn the Bf-109 and was capable of taking quite considerable battle damage. Other German fighter pilots confessed to sometimes sniggering when they heard the name '*Hurricane*', which in German sounds very much like '*Whore Barges*'. The criticism of the Hurricanes was largely unfair. They were simply, numerically overwhelmed at almost every turn.

On 15 May, the French Prime Minister Paul Reynaud (widely known as '*Mickey Mouse*') phoned Winston Churchill. Sobbing down the line, he admitted – just six days into the fight – that all was lost. The only

† Adolf Galland, who dismissed the Hurricane
† French Prime Minister, Paul Reynaud

REAL WAR/PHONEY WAR

the nation's army simply fled or else surrendered themselves before the advancing panzers. Britain was fighting France's war. Why, on the first day of the Blitzkrieg, French commanders had decided that their much vaunted air force was too exposed and had flown the bulk of it down south out of harm's way. Nevertheless, Churchill won and four more Hurricane squadrons were sent to France. That same day, Churchill flew to Paris under Hurricane escort. He found the French government and military in a state of utter collapse. There was much crying and throwing up of hands. The shrugging was over now. Officers and men ran to and forth, throwing piles of top secret papers into impromptu bonfires. Churchill felt so sorry for them that he left having promised yet another six squadrons of RAF fighters to the cause.

hope La Belle France had, he blubbered, was for Churchill to send over vast amounts of fresh British fighter aircraft to take on the Boche for him.

Churchill's instinct was to do what the French leader asked, but cooler minds resisted. Dowding immediately went into battle, bluntly saying that his fighter squadrons should be kept in reserve for when the Germans turned their attentions on Britain. We could win such a battle, he wrote in a memorandum, but not if we keep sending our fighters overseas in a futile bid to save France. Churchill pressed on with his desire to despatch more Hurricanes to France, but both the Air Secretary, Sir Archibald Sinclair, and the First Sea Lord, Sir Dudley Pound, turned on him and instead supported Dowding. It was pointed out that it seemed like the British were fighting on their own in France and that their fellow airmen in the French Air Force were very often avoiding the fight, while

✈ Sir Dudley Pound
✈ British Prime Minister, Winston Churchill

HAWKER HURRICANE THE HISTORY OF A LEGEND

'*Utter dejection was written on every face,*' Churchill recalled of the meeting – and there were whispers that – should they choose to – the Germans could be in Paris in just a few days.

Although his cabinet voted their approval, Churchill's opponents were not finished. It was Sir Cyril Newall, the Chief of the Air Staff, who suggested that, since the French airfields were so imperilled and in such bad repair, it would be wiser to operate new Hurricane squadrons from Britain and fly them over the Channel to fight each day and return each night to England. The idea was taken up, much to the relief of Dowding who was now at his wit's end trying to protect Fighter Command.

Back in France, fighter pilots were so exhausted by the constant fighting – five combat sorties a day was not uncommon – that they were reported to be landing whilst fast asleep. Meanwhile the French Air Force seemed to be doing nothing *'Demoralised'*, *'Shambles'* and *'disinterested'* were words being bandied about. RAF reinforcements made absolutely no difference. Over the weekend of 18-19 May, at least 68 Hurricanes were shot down. The Luftwaffe lost more aircraft, but then they had aircraft to spare.

On the following Monday, as the German panzers pushed for Calais, the surviving Hurricanes of the BEF were given the order to get out of France if they could and to return to England. Just 66 of the 261 Hurricanes made it home. 120 damaged Hurricane fighters had to be abandoned and destroyed on their French fields.

11 Group Hurricanes based in England tried their very best to provide adequate air cover for the hundreds of thousands of troops now waiting to be evacuated from Dunkirk. Some squadrons flew six sorties a day over the sea and pilots quickly became exhausted. Fuel limitations meant that they barely had time to offer air cover on the continent before having to return to refuel and rearm. Below them, on the beaches and quays of Dunkirk, the British Army could not understand why the RAF were not supplying better air cover and the resulting bitterness between the two was to fume for at least another two years, with reports of RAF pilots actually being physically assaulted by angry soldiers. In reality, 11 Group flew almost 2,800 sorties in support of the army at Dunkirk over nine desperate days, in the process shooting down 258 German fighters and bombers and damaging a further 119. 27 Hurricane

✚ L-R: Running repairs in France; Hurricane pre-checks; Dunkirk and the retreat

✚ Ground worker taking precautions in France

HURRICANE ACES

Flying Officer Edgar 'Cobber' Kain

Flying Hurricanes with No 73 stationed in France, 'Cobber' Kain was the RAF's first fighter ace of the Second World War'. A New Zealander by birth, Kain scored his first five kills during the so called 'Phoney War', bringing down two bombers and three fighters. During the Blitzkrieg which followed in May 1940, Kain scored another nine 'kills' in just 17 days. The combat took its toll and he was ordered to return to England to recuperate.

Before returning home, he said goodbye to his friends in the squadron by performing high speed, low level aerobatics over his airfield at Échemines. His Hurricane crashed and Kain died instantly.

pilots became 'Aces' during the air war over Dunkirk. They were joined in the action for the first time by Spitfires, but neither aircraft could deal with being massively outnumbered and the swarms of German fighters allocated to protect their bomber fleets often succeeded in keeping the RAF planes at a safe distance from them.

Meanwhile the remaining Hurricanes in France, those of the Advanced Air Striking Force were now retreating north, desperately trying to support Allied units still fighting the Germans while locating safe airfields to host them temporarily. To Dowding's fury, two fresh Hurricane squadrons were ordered to fly over the Channel to reinforce them. At a cabinet meeting, the Head of Fighter Command produced his statistics and told the cabinet that;

'If the present wastage continues for another fortnight, we shall not have a single Hurricane left in France or this country.'

The figures were stark and undeniable. The two new squadrons were never sent and, on 18 June, the last surviving Hurricanes in France were given orders to return home. 386 Hurricanes had been lost in the scant weeks following the Blitzkrieg – but they had claimed the lion's share of the 1,284 German aircraft shot down. As the last Hurricanes returned home, Churchill gave his 'Finest Hour' speech;

'…the battle of France is over. I expect the Battle of Britain is about to begin.'

☦ L-R: Hurricane ground crew in France; German troops check over abandoned vehicles at a Dunkirk beach

☦ RAF preparing to leave France

THE BATTLE OF BRITAIN

HAWKER HURRICANE THE HISTORY OF A LEGEND

It was inevitable that Germany should target Britain next, and its air force swiftly occupied captured airfields on the Channel and North Sea coasts. To support an amphibious invasion, head of the Luftwaffe Hermann Goering realised he must destroy the RAF first. He intended to gain victory by luring the RAF up for a decisive series of large-scale sky battles. In turn, Air Chief Marshal Dowding intended to achieve victory by doing the opposite of what Goering wanted. The RAF would fight many more small scale engagements, stretching out the conflict until winter came and a Nazi amphibious assault became impossible due to rough conditions at sea. The brunt of the fight would fall upon 11 Group, commanded by New Zealander Keith Park and responsible for the air defence of London and the south-east.

The Hurricane had proved an honourable foe over France but the Bf-109 pilots thought that they more than had their measure. The recently introduced Spitfire however they were much more wary of. Fighter Command's assessment concurred. The 29 available Hurricane Squadrons would be vectored against the 'easiest' targets – the German bomber formations. Spitfires would be reserved for tackling the bombers' fighter escorts, although still relatively rare. None were produced at the purpose built Spitfire factory in June, By contrast, Hawkers turned out 309 Hurricanes.

The Battle of Britain officially began on Wednesday 10 July 1940. The weather had improved and the Luftwaffe soon came out in force embarking on the first day of what they called *Kanalkampf* ('Channel Struggle'). Radar detected a large body of enemy

✠ German Luftwaffe chief, Hermann Goering; Keith Park
✕ Fleet of German Bf-109s; Hurricane being fuelled up
✠ Re-arming a Spitfire Mk I of 19 Squadron, between sorties

THE BATTLE OF BRITAIN

aircraft heading for a convoy codenamed *Bread* in the straits between Dover and Dungeness. RAF fighters were ordered to scramble in response – the Spitfires of 74 Squadron out of Hornchurch, 56 Squadron Hurricanes from North Weald, 111 Squadron Hurricanes from Croydon, 32 Squadron Hurricanes from Biggin Hill who were already out on patrol and – later on – six Spitfires from 64 Squadron out of Kenley. Facing them across the Channel were 24 Do-17s escorted by 30 Bf-110s and 20 or more Bf-109s.

111 Squadron hit the Dornier formation hard over Folkestone while the others went after their fighter escorts. The resulting massed air battle was a definite victory for the RAF. Two Dornier bombers were destroyed and no less than ten of their fighter escorts shot down. Only one ship in the convoy was sunk. However, the day did see the RAF's first official combat casualty of the Battle of Britain. P/O Thomas 'Peter' Higgs of 111 Squadron, a 23-year-old Hurricane pilot from Oldham, was blazing away on the tail of a Do-17 when he was bounced by a Bf-109. Surprised, he accidentally collided with the bomber and his fighter spun away out of control minus a wing. Higgs managed to bail out but died in the water. His body was washed up in Holland a month later.

On Thursday 11 July saw more convoy attacks despite the weather closing in again.

HAWKER HURRICANE THE HISTORY OF A LEGEND

The clashes of 10 and 11 July set the pattern for pretty much the rest of July. German bombers – in small or large numbers – would go after shipping in the Channel with or without fighter escorts. In turn radar would guide relatively small RAF fighter formations against them with varying degrees of success.

By the end of the month, it became clear to the German High Command that their tactics were not working. The RAF were simply not complying. No decisive pitched battles were being fought. Still, after the first month's combat, Goering found himself looking at 193 pilots and bomber crew confirmed dead and a further 302 missing. 216 Luftwaffe aircraft had been shot down during July. These figures did not augur well compared to RAF losses of 77 aircraft destroyed and 67 pilots and crew killed. He would have to try something else.

On 8 August, Goering was indeed planning for a final battle. He had summoned his top commanders to his palatial rural retreat – Karinhall – and, as he and his men played with Goering's prized toy train set, they worked out their plans for the 'Attack of the Eagles' – *Adlerangriff* – to commence with a first day's assault known as *Adlertag* (Eagle Day). If the RAF would not come out to fight en masse, the Luftwaffe would go to them. *Kanalkampf* was over. It was time to hit radar facilities, airfields and aircraft production plants. The RAF simply could not afford to lose these in Goering's thinking. They would have to come up and fight to protect them.

As the weather cleared somewhat on the morning of the 11th, a little after 10am the radar station at Ventnor on the Isle of Wight, reported a huge assembly of enemy aircraft over the Cherbourg Peninsula. Elements of 1, 87, 145, 152, 213 and 238 Squadrons were scrambled to meet it and to join the Spitfires of 609 Squadron who were already up on patrol. What they found as they closed with the enemy was nothing less than the largest single air raid launched

THE BATTLE OF BRITAIN

against Britain so far – a massed flotilla of 165 aircraft comprising 54 Ju-88s, 20 He-111s, 30 Bf-109s and 61 Bf-110s. The RAF fighters and the German fleet collided all along the Dorset coast between Weymouth and Swanage. It proved to be a brutal wake up call for both sides. By the end of the day, the Luftwaffe had lost 38 aircraft including 15 Bf-109s and 10 Bf-110s. The RAF lost 6 Spitfires and 21 Hurricanes, with an unusually high number of pilot fatalities. Large numbers of bombers had managed to get through while the British interceptors were entangled with the German escorts and they caused significant damage to facilities in Weymouth as well as to the Portland naval base.

On 12 August, the Luftwaffe went after Britain's radar defences, with Bf-110s temporarily knocking out three critical radar installations and then blitzing Portsmouth.

✠ L-R: Farmers working between raids; Hurricanes of 85 Squadron take to the skies; German He-111 bombers approach England's south coast

✠ Hurricanes ready to defend Britain

HURRICANE ACES

JAMES 'GINGER' LACEY

'I much preferred to kill someone without them even knowing I was there – the first indication he was being shot at was when bullets were coming out of his chest.'
– Ginger Lacey, BBC Radio Interview, 1978

Before the war, Yorkshireman 'Ginger' Lacey was training to be a pharmacist. As things turned out, instead he became the second highest scoring RAF fighter pilot of the Battle of Britain and one of Britain's top aces of the entire war.

Lacey joined No 501 Squadron flying Hurricanes in 1939 and scored his first 'Kills' on 13 May 1940 over France when he shot down both an He-111 and one of its Bf-109 escorts in a single sortie. Later that same time, he shot down a Bf-110. It was his first day in combat and, according to his daughter, *'when he landed no-one believed him.'*

Returning to Britain with 501, he scored his first 'kill' during the Battle of Britain – a Bf-109 – on 20 July. After racking up an impressive tally of confirmed kills and probables into August, he was shot down for a second time on 13 August by a gunner on an He-111. He was shot down for a third time on 30 August but managed to glide his stricken Hurricane to his home airfield at Gravesend. The very next day he destroyed a Bf-109.

Through September, Sergeant Pilot Lacey 'bagged' two more Bf-109s on 2 September and two further Bf-109s on 5 September. On 13 September, he engaged a flight of He-111s that had just bombed Buckingham Palace, shooting down one of them after being injured by one of its gunners (for shooting down this particular bomber, Lacey was given a most unusual award – the first parachute to be made in Australia together with a silk scarf embroidered with the names of hundreds of girls who worked at the parachute factory). Two days later, during some of the heaviest aerial combat seen during the Battle of Britain, he scored an incredible four 'kills' in a single sortie – bringing down an He-111 and 3 Bf-109s.

Having already been awarded a Distinguished Flying Medal for his bravery in combat back in August, Lacey received a second DFM in November 1940. By this time his tally of enemy aircraft had risen to 23 – 18 of those during the Battle of Britain.

'I was shot down nine times in 16 weeks. Twice I got out with my aeroplane burning from end to end, once with no tail on it.'
– Ginger Lacey, BBC Radio Interview, 1978

Lacey survived the war with 28 confirmed 'kills' and didn't leave the RAF until 1967 at the rank of Squadron Leader. In civilian life, he gave flying lessons and ran an air freight business. He also acted as technical advisor on the 1969 film *'Battle of Britain'* and the director, Guy Hamilton, later commented;

'I entrusted Ginger Lacey to be my main adviser. He really was a good chap, totally invaluable.'

James 'Ginger' Lacey died on 30 May 1989. He was 72. His childhood home, which was to bear a blue plaque in his honour, had since been demolished. That plaque is instead displayed in the entrance to the Aldi supermarket that now occupies the site. Aldi is German.

13 August was the day Goering had designated as Adlertag – Eagle Day. Unfortunately for him, the day broke with thick low cloud and fog and early sorties were cancelled. Nevertheless, by 1pm, the word had spread that Adlertag was now on again. The Germans threw their largest attack formation so far into the air. Leading the charge were 30 Bf-109s with 52 Stukas immediately behind and another formation of Bf-109 escorts to the rear. To the east flew 120 Ju-88s and 30 Bf-110s. To the West were 27 Stukas. Advanced parties of Bf-109s swept over Portland, acting as bait and hoping to draw away RAF fighter formations but didn't succeed. Instead, 80 Hurricanes and Spitfires rose up to meet the main formation.

Luftwaffe bombers punched their way through to Southampton docks causing widespread damage, the Luftwaffe ended the day estimating that they had destroyed 84 Spitfires and Hurricanes. They were wrong. In reality, they had only claimed 13 RAF fighters and all but three of the pilots shot down had survived. In turn, the RAF thought they had accounted for some 64 Luftwaffe fighters and bombers. The real figure was 47. In numbers terms, Adlertag had turned out to be – if anything – a victory for the RAF, but no-one knew it at the time. Goering wanted to keep the pressure up on the following day but was confounded by more poor weather conditions.

HAWKER HURRICANE THE HISTORY OF A LEGEND

more. As night fell, the Germans tallied up their losses. They had flown 2,199 sorties compared with the RAF's 974 and lost seventy five aircraft in a single day, compared with RAF losses of just 30 with 13 pilots killed. Churchill would call 15 August *'one of the greatest days in history.'* Henceforth, the Germans would refer to it as *'der schwarze Donnerstag'*. Black Thursday.

Luftwaffe massed raids continued the next day, targeting airfields and radar stations. At one point in the day every single fighter squadron in the Group had been committed – and yet more German raiders were inbound according to the radar. Squadrons were being called upon to fly four or even five separate combat missions over the course of a single day, racing home to refuel and rearm between sorties.

By 15 August, Goering understood that Adlertag had not been the success he had been counting on – or the one he had promised the Fuhrer. As he complained to senior commanders gathered at his home, 800 Luftwaffe bombers escorted by 1,000 fighters took off for England, comprising the largest raid of the war so far. Maybe this time would be different. They inflicted serious damage on the airfields at Hawkinge and Lympne and radar stations at Dover, Rye and Foreness. Later that day, more attacks were launched against Hawkinge aerodrome, Maidstone, Dover, Middle Wallop, Portland, Worthy Down, Rye and the radar station at Foreness once

✢ A German Junkers Ju-88 drops its load
✗ Dornier Do-17s sweep over Britain
✢ Hurricane Mk I of 85 Squadron patrolling the sky

THE BATTLE OF BRITAIN

On Sunday 18 August the Luftwaffe came again – and in almost overwhelming force. By lunchtime, radar had picked up huge German formations assembling across the Channel. Every single Squadron in 11 Group was put on alert. The fighting started badly, with Bf-109s ambushing 501 Squadron over Canterbury and shooting down four Hurricanes. Dorniers, flying at almost treetop height, surprised the defences at Biggin Hill but were then set upon by the Hurricanes of 32 and 610 Squadron as they accidentally rained bombs down on a local golf course. Only two of the Do-17s made it home to France. More low flying Dorniers struck Kenley, dropping over a hundred bombs and destroying ten Hurricanes on the ground while putting every hangar out of action. West Malling and Croydon were also hit. More raids followed.

As the last German aircraft touched down on their bases that evening, it became apparent that the RAF had won the day again. It was believed that 140 Luftwaffe aircraft had been accounted for. The

truth was closer to 70. The RAF for their part had lost just over 30 planes in combat, not counting those destroyed on the ground by air raids. Only 10 pilots had been killed. Goering was incandescent with fury. He called his fighter pilots cowards and sacked some of his officers, replacing them with men he felt had more of a killer instinct.

On the British airfields now, the pilots were facing near-exhaustion. They could be called upon to fly

possibly five combat sorties every single day – and flying seven sorties was far from unheard of. They had to be on call from 3am every single morning. The pilots of 32 Squadron at Biggin Hill took to sleeping out under the wings of their Hurricanes using their parachutes as pillows, while 151 Squadron pilots at Rochford took to sleeping in the cockpits of their already fuelled up and armed fighters.

It was not until Saturday 24 August that the Germans returned with any real force. Over the course of the day, they sent more than 500 aircraft over the Channel in six concerted waves. Raiders returned by night looking for the Thames Haven oil storage depot and the Short's aircraft factory in Rochester. Navigation proved a problem though and ten aircraft dropped their bombs over civilian London, starting fires from Bethnal Green to Oxford Street. Nine Londoners died. Bombing London was strictly forbidden by Hitler. The bombing of London, however, was to have huge consequences – and change the direction of the entire Battle of Britain.

Late August saw Goering having to resort to dummy bombing raids shadowed by vast numbers of fighters just ready to pounce. The RAF by and large did not

rise to the bait. Goering's lust for one decisive battle continued to frustrate him.

Galled by their failure to lure the RAF into the trap, the Luftwaffe threw almost everything they had at South East England on Friday 30 August. Three waves of enemy aircraft swept in. The first was comprised entirely of fighters looking for a dogfight. The second, half an hour later, mixed 70 He-111 and Do-17 bombers with 90 fighter escorts. They were hit first by 11 Hurricanes of 85 Squadron which helped to break up their formation. Radar operators were stunned to

✈ ABOVE: The London bombing had started, albeit by error
L-R: Wrens assist in Hurricane repairs; More Hurricanes ready to take on the fight; Do-17 formation over the south of Britain

✈ A Heinkel He-111 flies over East London INSET: Hurricane promo poster

see their screens almost filled with aircraft signals. The skies over South East England were packed with aircraft formations and wheeling dogfights. By noon everything 11 Group had was in the air.

The Luftwaffe started early on the morning of 31 August, striking against airfields and radar installations alike. The hits on the airfields were starting to cripple the fighter squadrons on the ground, disrupting them and depriving them of essential facilities. At the same time, the massed fighter escorts were proving effective cover for the bombers, On 31 August the Luftwaffe still lost 39 aircraft – but the RAF lost 37 of their own, with 13 pilots killed. The battle was now turning in favour of the Germans…

The accidental bombing of civilian areas of London back on the night of the 24/25 gave Winston Churchill all the justification he needed to lash out against the capital of Nazi Germany. On the night of 25/26 August, 95 Hampdens and Whitleys from RAF Bomber Command were despatched to bomb Berlin's Tempelhof Airport and weapons factories located to the north of the city, Few Berliners were killed in the raids and damage was only slight, but both Goering and Hitler were beside themselves with fury at this affront to the Reich. Hitler's rage and sense of insult

✈ **ABOVE:** Handley Page Hampden L-R: Hurricane at Castle Camps during the Battle of Britain; Heinkel Tempelhof Airport in Berlin; Downed German aircraft inspection

✈ Dogfight trails over St Paul's Cathedral London

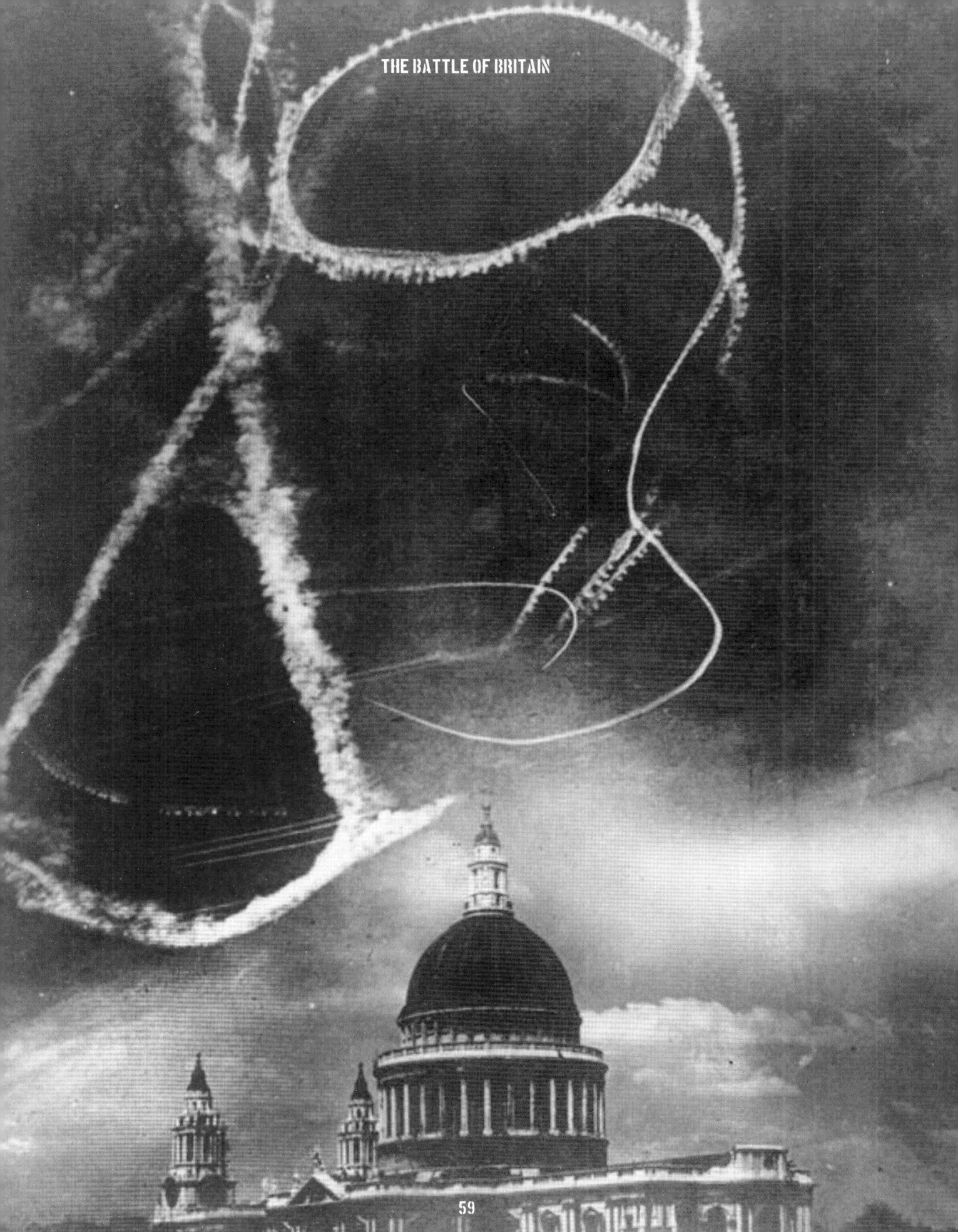

THE BATTLE OF BRITAIN

303 SQUADRON POLISH

✞ Members of 303 Squadron in jubilant mood after Battle of Britain combat

Flying Hurricanes out of Northolt, 303 (Polish) Squadron soon earned a proud reputation as one of the elite RAF Squadrons during the Battle of Britain.

However, they were very late to enter the battle, due to doubts held about them by RAF commanders. They were told that they could not fly until they could speak English and were given strange training exercises like cycling in strict formation around Ruislip with radio transmitters attached to their bikes, practising following orders.

They were led by a short, English ex-stockbroker called Squadron Leader Ronald 'Boozy' Kellett, who delighted his Polish comrades with his eccentric ways and his love of driving everywhere in his own Rolls Royce. His nickname might also offer another reason for which he was respected. A genuinely strange man, in his way as undisciplined as his Polish charges, he was known to try and train pilots by pointing to a Hurricane and then flapping his arms furiously and yelling, *'fly!'*

The squadron scored its first 'kill' before even becoming operational. On 30 August, whilst on a training flight, Flight officer Ludwik Paszkiewicz spotted a Bf-110. He asked permission to engage and when he got no reply, just went ahead and did it anyway, closing to point blank range before blasting it from the skies. On his return to Northolt, Paszkiewicz was given a public roasting in front of his comrades by Kellett, but then heartily congratulated in private by him. Kellett then rang his superiors and requested 303 Squadron become officially operational. They were given operational status from the very next day – and scored 4 confirmed 'kills' and two further probable.

303 Squadrons quickly proved their proficiency in air combat. On 5 September, they shot down three Heinkel-111s and five of their Bf-109 fighter escorts for the loss of just one aircraft. On 11 September, 303 Squadron shot down three Do-17 and four He-111 bombers, as well as two Bf-110s and a Bf-109 for the loss of two aircraft. They were also in the thick of the fight on the crucial day of 15 September. Mid-morning, the Poles shot down two Do-17s, a Bf-110 and six Bf-109 escorts over London – and returned in the afternoon to claim another four kills.

Praise flooded in for the Polish contribution to the battle, even from Downing Street. Squadron Leader Kellett replied to Churchill's warm words to the effect that talk was fine but they'd rather have a crate of whiskey thank you very much. They got it. A local school in Ruislip had a collection for the airmen and all clubbed in to buy them 450 cigarettes. Another school in Glasgow proudly sent them a postal order for ten shillings. On 26 September King George VI came to pay a personal visit. While he was around, there was a squadron scramble and the king was almost knocked off his feet as the Poles lost all interest in His Majesty and ran for their fighters.

Skilful and lethally aggressive, the pilots of 303 Hurricane Squadron went on to claim a total of 126 kills during the Battle of Britain – the highest score of any of the 66 Squadrons which participated – even though they had not become operational until the dying days of August 1940. They also set the record for the highest number of kills in a single month (more than double that of the nearest squadron) and for the most kills for the fewest losses. Their desire to kill Germans at any price is amply illustrated by the actions of Sergeant Stanislaw Karubin who, whilst chasing at Bf-109 ct treetop height on 2 September, suddenly found his guns had run dry. Undeterred, he flew directly over the top of the Bf-109, matched it for speed and then gradually dropped his height, forcing the German pilot right into the ground.

✠ 303 Squadron pilots with their puppy dog mascot

to start rearranging his fighting assets. The Germans were suffering too. During July and August, they had lost some 800 fighters and bombers and – on the day of Sunday 1 September they had less than 600 frontline serviceable aircraft to put into the air.

Into September, Goering continued to try and tempt the RAF into a massed battle, now flying huge swarms of fighters mimicking the behaviour of bombers to fool radar. On 3 September, he finally switched tactics – a grave error. The Luftwaffe should now turn its attention to London, as ordered by the Fuhrer himself in retaliation for the RAF raids on Berlin. This, surely, would lure the RAF up for a final, decisive showdown, Goering reasoned.

It took the Luftwaffe another few days to gear up for attacks on London. In the meantime they continued with the strategy of targeting the airfields. 7 September was a beautiful sunny day. After a morning spent touring his airfields, Goering sat down on the cliffs of Cap Gris Nez overlooking the Channel and unpacked a large and lavish picnic. As he tucked in, he was quite convinced that today he would see the end of the RAF. By 4pm, a force of almost 1,000 Luftwaffe fighters and bombers were assembling in the air above – 348 bombers with 617 fighter escorts. It was the greatest flotilla the Luftwaffe had ever put into the air, a formation two miles high and spread across almost 800 square miles.

would gnaw at him until, on 4 September he would announce to those assembled at a rally in Berlin' *we will raise their cities to the ground'*.

The large-scale air battles and constant pummelling of aerodromes during late August had left the RAF in a parlous state. By 1 September, Dowding was forced

Stunned by the radar signals and reports of *'many hundreds of aircraft'* by the Observer Corps, the RAF initially threw up 11 squadrons and readied all the rest. 10 and 12 Group were put on standby. It was a complete shock when bombs began falling on the

✈ Heinkel He-111s ready to despatch their bombs; Firemen wrestle with their hose

✈ Smoke bellows from the London landscape around the Tower Bridge area

city of London. Below, the air raid sirens sounded at close on quarter to five and the Londoners heading for the shelters described the sky as being *'black with aircraft'*. At 5pm, Dorniers and Heinkels released their bombs and incendiaries onto the East End from three miles up. Their initial bomb run was virtually unchallenged except for anti-aircraft fire from the London batteries – but then RAF squadrons began to arrive – 23 of them – diving on both the returning bomber formations and the new ones coming in. Bf-109 escorts slammed into them in turn. Those not in shelters were treated to the sight of what was effectively a thousand aircraft dogfight high above their heads. The Bf-109s had to break off first as their fuel ran low and the bombers were left to fight their way home, harassed by the RAF fighters until their tanks too ran low.

The 'all clear' sounded at 6.15 and civilians emerged from their shelters to find the East End aflame. The bombing had been ferociously accurate. Many docklands warehouses contained combustible produce including the Surrey Docks which held almost 2 million tons of wood supplies. Everything seemed to be on fire and the firemen initially tasked with dealing with the conflagration suffered terrible losses as they were overwhelmed, choked, suffocated and then engulfed in the firestorm. Extinguishing the blaze was impossible – it stretched for nine whole miles along the banks of the Thames. Whole streets in the East End simply ceased to exist.

The next day, Park saw the devastation for himself as he flew over the stricken city in his personal Hurricane. He was horrified – but he also grasped right there and then that the Germans had made a potentially fatal tactical mistake. By switching their attention to London, they would be giving the RAF airfields a vital breathing space in which to sort themselves out.

The air defence of London had not gone well, however. Only 38 German fighters and bombers had been shot down and in doing so the RAF had lost 28 aircraft of its own. The bombers returned at 7.30pm on the 8th. As they came in from the east following the Thames, the pilots tuned their cockpit radios into the BBC and listened to dance music as they bombed.

On Monday 9th, 605 and 607 squadrons managed to disrupt and break up a raid by Do-17s and He-111s heading for London. 19, 242 and 310 squadrons of 12 Group joined in the defence of London, deciding themselves to get involved rather than stick to their assigned role of airfield defence and a large German force was met with unexpected strength over South London. Faced with determined resistance by 253 Squadron Hurricanes and 303 (Polish), many bombers just dropped their loads early and fled for home. They had been expecting a clear run. The RAF was quickly re-gearing to meet the threat to London. Most of the day's raids were driven off and 28 Luftwaffe aircraft downed at a cost of 21 RAF fighters.

HAWKER HURRICANE THE HISTORY OF A LEGEND

By night, the Luftwaffe came again. Under cover of darkness, it was almost impossible to find let alone stop the bomber fleets. Another 370 Londoners were killed in the night raids as Madame Tussaud's, the Natural History Museum and the Royal Courts of Justice all burned. The 10th was a similar story. Two more large scale daylight raids were launched against London on 11 September. German losses amounted to around 25 aircraft but the RAF lost 29 fighters in intercepting the raids. That night 200 Luftwaffe aircraft returned to pound London once more, while others struck Liverpool. The raids continued until inclement weather offered both sides a respite.

The weather however was perfect on the morning of Sunday 15 September. Everyone knew this meant the Germans would be coming in force. At around 10.45am, radar reports started coming in of a massive German force assembling over the Pas de Calais – a force two miles wide and nine thousand feet thick in the air. It comprised a hundred Dorniers escorted by two hundred Bf-109 escorts. The target was going to be London, All 11 Group squadrons were readied in a matter of minutes and 10 and 12 Group notified of the need to help. Park had devised a new tactic. He knew he had no hope of stopping the force getting through but – if his fighters could just keep the escorts busy and entangled over southern England, they would use up their fuel quicker and would have to turn for home earlier. That meant the bombers would be unescorted – and more vulnerable – for longer.

Spitfires of 72 and 92 Squadrons hit the Luftwaffe formation over Maidstone. The Bf-109s went after them, only to be bounced by 603 Squadron. More Bf-109s got sucked in to aid their fellow Messerschmitts, creating a hole through which 253 and 501 fighters managed to get at the Dorniers. More and more RAF squadrons piled into the epic aerial battle as the formation progressed over Kent, until the Germans found themselves under attack by 250 fighters from 23 different squadrons. A further six squadrons joined the fight as the bomber fleet reached the suburbs of south London, then five squadrons of the Duxford Wing of 12 Group came diving down on the Dorniers just as they were dropping their bombs. The bomber formations fell apart, individual planes wheeling and accelerating everywhere in a desperate attempt to escape. Having broken formation, there was little their escorts could do to protect them. RAF fighters harassed them all the way south back over the Channel as well as up the length of the Thames into the North Sea.

As the RAF squadrons raced home to land, refuel and rearm, reports came in of another colossal German aerial armada sweeping in – in three waves – a total of 150 bombers escorted this time by 400

✈ L-R: Bf-109s during a raid; The mangled mess of a downed German fighter; Daunting sight of a German Bf-110

✈ Spitfires of 610 Squadron flying in formation over Britain

THE BATTLE OF BRITAIN

fighters. They crossed the coast near Dungeness at 2.15pm, spread across ten full miles of sky. Spitfires from 41 and 603 Squadrons hit them over Romney Marshes and managed to shoot down two of the Dorniers before the Messerschmitt escorts fell on them. 73 Squadron were next, destroying three bombers off Maidstone. Spitfires from 66 and 72 Squadrons intercepted them over Dartford, quickly reinforced by four more squadrons. Hurricanes of 303 (Polish) hit bombers and their Bf-110 fighters over Gravesend. 213 and 607 Squadron Hurricanes out of Tangmere punched into the bombers above the borders of Surrey, shooting down several in their initial attack. Wherever the bombers flew, they flew into more trouble.

The nerve of the Luftwaffe bomber pilots began to crack under the sheer weight of the assaults on them. The formations teetered. Some individual bombers broke formation and fled. Others dumped their bombs. Those who held their nerve faced a wall of flak from London's recently improved AA defences as they reached the outskirts of the city. After negotiating that, they suddenly found themselves facing 15 more squadrons of RAF fighters, comprising a daunting Big Wing of five 12 Group Squadrons and ten 11 Group squadrons. It was a truly terrifying sight. The German bomber formations – suddenly faced with a sky full of fresh hostile fighters – lost what was left of their discipline and their formations fell apart. They dropped their bombs pretty much just anywhere and then tried to flee for home. They were harassed by fighters every single step of the way.

Over in France, the sight of one damaged bomber after another limping its way home and then spilling out wounded and dying air crew told its own story. The RAF were far from finished and something pretty

HAWKER HURRICANE THE HISTORY OF A LEGEND

much decisive had just happened. Back across the Channel, the RAF pilots and their commanders knew that he had achieved something incredible. There were reports of 183 German aircraft shot down. It wasn't true – the real number was 56 – but it felt to the fighter pilots as if they had decimated the Luftwaffe that day. RAF losses stood at just 26 – and a full half of the pilots shot down had survived.

October came. By the 12th Hitler had agreed to abandon his invasion plans until at least the following year, while Goering felt compelled to offer his demoralised fighter pilots free skiing holidays. The end of October merely confirmed that the RAF were in charge now. The month had seen 379 Luftwaffe aircraft shot down for the loss of 185 RAF fighters. Today 31 October is officially regarded as the last day of the Battle of Britain. It was quiet due to rain on that day, and there were no losses.

The air battle had been very evenly matched. The RAF had lost 1,087 planes and 481 pilots. 357 Spitfires had been shot down, together with 601 Hurricanes. The Luftwaffe had lost 1,652 aircraft and just over 3,000 pilots and air crew. Of the aircraft lost 778 were bombers. 533 Bf-109s were shot down.

The Battle of Britain might be over, but the Blitz would continue with full ferocity. From the end of October, the Luftwaffe concentrated on its night raids, and the bombing of London would continue almost every single night until May 1941, when the bombers were called away to prepare for the invasion of Russia.

SPITFIRE VERSUS HURRICANE

In total, it's agreed that the Luftwaffe suffered losses of 2,739 aircraft. Hurricanes accounted for 55% of them. Spitfires killed 42%. However, the Spitfire captured the public imagination – and the hearts of the British people – in a way the Hurricane never could achieve. In the dark and desperate days of June 1940, the Hurricane was what Britain already had. The Spitfire was presented as a newly arrived super weapon, the aircraft which would save the day and the nation. It gave fresh hope.

The Hurricane was simply under appreciated. The British people, via the Spitfire Fund raised some £13 million (around £670 million today) for Spitfires. No one was out collecting for Hurricanes. For once, the British people's traditional love of the underdog seems to have gone AWOL.

The Battle of Britain was barely over before analysts set about working out how the aircraft of Fighter Command had performed. What they really wanted to know was whether the Hurricane or the Spitfire had performed best against their Luftwaffe opposition. More Hurricanes had flown and fought in the battle than Spitfires – and the two aircraft were often deployed in quite different ways – but, allowing for this, the maths quickly proved what everyone already knew. The Spitfire was clearly the better fighter plane.

It was calculated that the average Spitfire survived combat for a third longer than a Hurricane. It was also noted that, for every 23 Luftwaffe aircraft accounted for by Hurricanes, Spitfires managed to bring down 27 despite often being vectored against fighters rather than bombers.

On 14 October 1940, it was noted in an Air Staff meeting that '...*the Hurricane is found not a match for the Messerschmitt. The Spitfire retains its superiority.*'

Production and development would now concentrate on the Spitfire.

✙ The Spitfire was certainly proving to be a worthy and capable aircraft

✈ 85 Squadron Hurricanes

HURRICANE ACES

ARCHIE MCKELLAR

Archie McKellar, a native Scot, is credited as the highest-scoring Hurricane pilot of the Battle of Britain. Before the war, he worked as a stockbroker but disliked the job so intensely that he chose to become a plasterer instead.

Serving as the Squadron Leader of 605 Squadron, he shot down 21 enemy aircraft. His men called him 'Shrimp' as he stood just 5 foot three inches tall.

He was only one of two RAF Battle of Britain pilots credited with becoming '*an ace in a day*' when, on 7 October 1940, he shot down five Bf-109s. McKellar died in a dogfight with a large number of Bf-109s on 1 November 1940, having been awarded the DFC and Bar for his valour in combat. A posthumous DSO followed.

INTO THE BLITZ

HAWKER HURRICANE THE HISTORY OF A LEGEND

HURRICANE BY NIGHT

The Battle of Britain was over, and the air war shifted gear. Now the Luftwaffe came mostly by night, its bomber fleets unleashing the Blitz upon British cities. No-one in the British High Command had adequately foreseen this and as a result night fighters were virtually non-existent when the need came.

In a crisis, it was decided that Hurricanes should be pressed into service to protect British night skies. Although Spitfires would serve as night fighters too, it was thought that the Hurricane would offer an advantage over its more celebrated rival. It had a

✈ Coventry sustained much bomb damage

wider wheel track, which meant it was safer to land in at night. The Hurricane's cockpit also offer better all-round visibility than the Spitfire, a real consideration when trying to spot an enemy bomber under the most difficult of circumstances. Five Hurricane Squadrons were rapidly painted jet black and assigned to the task of stopping the Blitz.

It didn't work. Ground controllers using radar reports could guide the Hurricanes into the general vicinity of enemy bomber fleets, but once in the area, pilots found locating individual aircraft an impossible task in the darkness. On 14 November 1940, fleets of Luftwaffe bombers over Coventry turned the city into a vast raging inferno. There were 449 German

HURRICANE ACES

FLIGHT LIEUTENANT RICHARD STEVENS – THE SPIRIT OF VENGEANCE

'My God he was a bloody killer. His hatred was unbelievable.'
– John Ellacombe, 151 Squadron

Before the war, Richard 'Steve' Stevens was a commercial pilot. When war came, he served with Fighter Command flying Hurricanes but those who flew alongside him said that Steven's personality changed completely the day his wife and children were killed in a bombing raid on Manchester. He became an utterly ruthless killer.

Flying with 151 Squadron out of RAF Wittering near Peterborough on the night of 15-16 January 1941, he scored his first kill – a Dornier Do-17 bomber, closing and opening fire when he was just 25 yards from the bomber. Stevens' manoeuvres to get into position were so violent that he actually cracked the fuselage of his Hurricane – and wrecked his eardrums. As one of his fellow pilots recalled, he had been so close to the Dornier when he opened fire that the 'blood and guts' of its rear gunner were all over his Hurricane's wings when he returned to base. His Ground crew were forbidden to wash it clean, Nine more night kills followed by June 1941. On one occasion, a fellow pilot recalled, Stevens asked for a lift to the crash site of a Heinkel He-III he had recently despatched. On reaching the wrecked bomber, it was found that the impact of the crash had driven the rear gunner's machine gun deep into his chest. While his fellow pilot threw up, Stevens wrenched the gun out of the German's ribcage and went off to rinse it clean in a nearby river. He then kept it as a trophy. *'He hated bloody Germans,'* his compatriot later explained.

Stevens' fellow pilots started to realise he was a man possessed and filled with a singular hatred. He kept himself to himself, living the life of an obsessed loner. All he wanted to do was to fly and to kill. He was an excellent marksman, and his ability to find an enemy aircraft in a pitch black sky was uncanny.

Eventually, he was allowed to go his own way. He would take off in his black Hurricane, switch his R/T off and simply go hunting. Between June and December 1941, he accounted for four more German intruders. On the 12 December, he took his black Hurricane up on one final mission over Holland and simply never returned.

HAWKER HURRICANE THE HISTORY OF A LEGEND

bombers over the city that night. Hurricane night fighters swept in to exact revenge, but couldn't make contact with a single one. *'None of us saw a thing,'* confessed Douglas Bader, then Squadron Leader of 242 Squadron. Another Hurricane pilot on a different intercept complained that his squadron *'stooged blindly around.'* He added bitterly, *'we had no success whatsoever'*. Chillingly, some Hurricane night fighter pilots said that they were more likely to locate a German intruder not by sight but by their aircraft suddenly being hit by the swash of turbulence as one passed close by in the darkness. In January 1941, the whole contingent of RAF night fighters managed to shoot down just three enemy aircraft.

Dowding recognised early on that the only true solution to the problem was to fit night fighters with their own onboard radar. Experiments began, but favoured the two engined Beaufighter because it could carry the relatively heavy equipment needed. It was not until 1942 that the first Hurricane night fighters could be made to accommodate radar.

CIRCUS, RAMROD, RODEO, RHUBARB

In the first half of 1941, the Blitz eased up and German bombers increasingly disappeared from British skies. Hitler needed them for his forthcoming Operation Barbarossa – the invasion of Soviet Russia. The pressure was off – and now the RAF could actually go on the offensive over occupied Europe, with raids variously codenamed Circuses, Rodeos, Rhubarbs and Ramrods.

HURRICANE ACES

ROBERT STANFORD TUCK

'I've just been bloody lucky, that's all.'
– Robert Stanford Tuck

Londoner Robert Stanford Tuck started his impressive tally of kills on Spitfires until 11 September 1940 when he was promoted to Squadron Leader and given 257 Squadron, which flew Hurricanes. Tuck racked up a total of seven 'kills', plus four probables and two damaged. Although initially critical of the Hurricane, he later wrote of the aircraft;

'It was a remarkably good gun platform; very steady when you opened fire....It was very easy to fly; had no vices, and would take a great deal of punishment and bring you back home... So it was a very fine aircraft for fighter vs. bomber work.'

His enthusiasm for the Hurricane only increased when his squadron converted from Mk Is to Mk IICs with their four 20mm cannon and he became a vocal advocate for them.

In January 1941, Tuck was awarded the Distinguished Service Order (DSO), to add to the two Distinguished Flying Crosses he had won during the fighting over France and the Battle of Britain. A third DFC followed in March 1941.

Post war, with a final tally of 29 kills (and a further two shared), Tuck was promoted to temporary Wing Commander in 1946. He retired from the RAF in May 1949 and passed away in 1987 aged 70.

Circuses were bombing raids accompanied by fighter escorts like Spitfires and Hurricanes up to 16 squadrons strong, the intent being to lure large numbers of German fighters into the air and inflict maximum casualties. With Ramrods, the composition of fighter and bomber elements was the same, but the bombing damage was considered more important than the fighter shoot-downs. Rodeos were fighter-only sorties with Spitfires and Hurricanes

✈ L-R: Bristol Type 156 Beaufighter; Hurricane painted in night fighter colours

✈ Douglas Bader

✈ Spitfires having taken off from Biggin Hill

HURRICANE ACES

KAREL KUTTELWASCHER

'The greatest figures among our fighter pilots.'
— Air Vice-Marshal Karel Janoušek

Czechoslovakian by birth, Kuttelwascher joined the Czechoslovak Air Force in 1934. When the Nazis invaded in March 1939, he hid on board a coal train and escaped to France where he joined the French Air Force. During the Blitzkrieg he shot down at least two enemy aircraft before escaping to Britain.

Recognised as a pilot, Kuttelwascher was invited to join the RAF as a flight sergeant and retrained on Hurricanes. He joined No.1 Hurricane Squadron on 1 October 1940 and took part in the last few weeks of the Battle of Britain.

His first confirmed 'kill' with the RAF came while his squadron was flying combat missions over occupied France. On 8 April 1941, he brought down a Bf-109. A second followed in May and a third in June. Hunting now by the light of a full moon, 1 Squadron switched to night fighting with their new all-black Hurricane Mk IICs, tracking down and pouncing on enemy aircraft as they returned to base. Kuttelwascher excelled in this role. Flying his Hurricane nicknamed *'Night Reaper'*, he became the RAF's most successful night intruder.

In just three months, Kuttelwascher destroyed 15 enemy planes and received the Distinguished Flying Cross. Today he is officially credited with 18 'kills'.

prowling for prey to shoot up on the ground and Rhubarbs were another version of this tactic, setting Allied fighters against both air and ground targets.

Casualties were high, even for Spitfires, while slower, more vulnerable Hurricanes only demonstrated by now that they were well on the way to obsolescence. They were quickly consigned to ground attack missions only. Over 300 RAF pilots were lost on these sorties.

CHANNEL STOP

'E-boats, if you hit them, burned up very nicely.'
– Denys, Squadron Leader. 615 Squadron

Hurricanes suffered fighting other fighters, but proved more successful on Channel Stop raids against enemy shipping. These attacks were usually conducted by flights of eight Hurricanes split into two low level waves

diving in with mgs and cannons. They became even more effective when rockets were introduced.

DIEPPE

By August 1942, despite having been in the war for less than a year, the Americans were already pressing for the Allies to hop over the Channel and invade Europe. The British were sceptical to the point of being frightened by the naive eagerness of their new friends and kept trying to find excuses and distractions.

D-Day would have to come eventually though – but what would it be like?

On 19 August 1942, the Allies launched a large scale raid on the fortified coastal town of Dieppe in France, with some 6,000 men (mostly Canadian) hitting the beaches supported by 237 ships and hundreds of aircraft. It was also to be the last major concerted fight for the Hurricane in the European theatre.

Eight squadrons of Hurricanes took part in support of the beach landings during 'Operation Jubilee', ranging from a few 'vintage' Mk 1As to cannon-wielding Mk IICs. Aware that any Hurricanes would be reduced to mincemeat by the new Fw-190s in the area, the Hurricanes were instead ordered to attack German ground positions and to avoid dogfighting. That was to be left to the Spitfires.

The Hurricanes did a creditable job in providing ground support but their 250 and even 500lb bombs proved insufficiently powerful to do much damage to German heavy fortifications, while seemingly everything else around them went wrong. Allied tanks became stuck on the beaches and troops coming ashore were held off by fierce enemy resistance, In just a few hours over half the men would be killed or taken prisoner. In the skies, 106 RAF planes were shot down as the latest Spitfire IX proved little match for the Fw-190s. Twenty Hurricanes were shot down during the operation and fifteen pilots killed or captured.

✈ L-R: Dieppe raid; Hurricane dogfighting; Landing craft en route to Dieppe

✈ German Fw-190

SIGNIFICANT VERSIONS AND VARIANTS

HAWKER HURRICANE THE HISTORY OF A LEGEND

Mk I

The first mark of Hurricane, produced between 1937 and 1939, varied widely because of improvements made during production runs. It might have a two or three bladed propeller or be powered by a Merlin II or III engine. Standard armament was eight Browning .303 machine guns. From early 1940, the Revised version featured a constant speed propeller. In total 4,200 were built.

Mk II

The summer of 1940 saw Hurricane Mk II's fitted with the Merlin XX engine with two-speed supercharger. Mk IIB's were now fitted with twelve Browning machine guns and racks for optional drop tanks. Drop tanks were also an option for the Mark IIC, which could also carry a par of underwing bombs, reflecting the change in the Hurricane's status from a fighter to a ground attack aircraft. The machine guns were replaced by a pair of 20mm Hispano cannon on each wing. The 'D' variant sported twin 40mm Rolls-Royce/Vickers anti-tank cannon, one under each wing. 296 'Flying can-openers' were built and went into action against enemy vehicles from heights as low as twenty feet.

SIGNIFICANT VERSIONS AND VARIANTS

Mk III

A unsuccessful test version of the Hurricane, powered by a Merlin engine built by US manufacturers Packard Bell.

Mk IV

524 Mk IV's were built from December 1942 onwards. They featured a new 'universal wing' which enabled the Hurricane to carry a wider array of armaments, bombs, smoke cannisters, drop tanks etc.

Mk V

The last official version, used to test more powerful engines. Only one was ever built.

X

Canadian-built variants were always designated with an X.

✈ Side profile of Hurricane
✈ T-B: Hurricane Mk II; Mk IV; Mk X
✈ Hurricane Mk I of 315 Squadron

HAWKER HURRICANE THE HISTORY OF A LEGEND

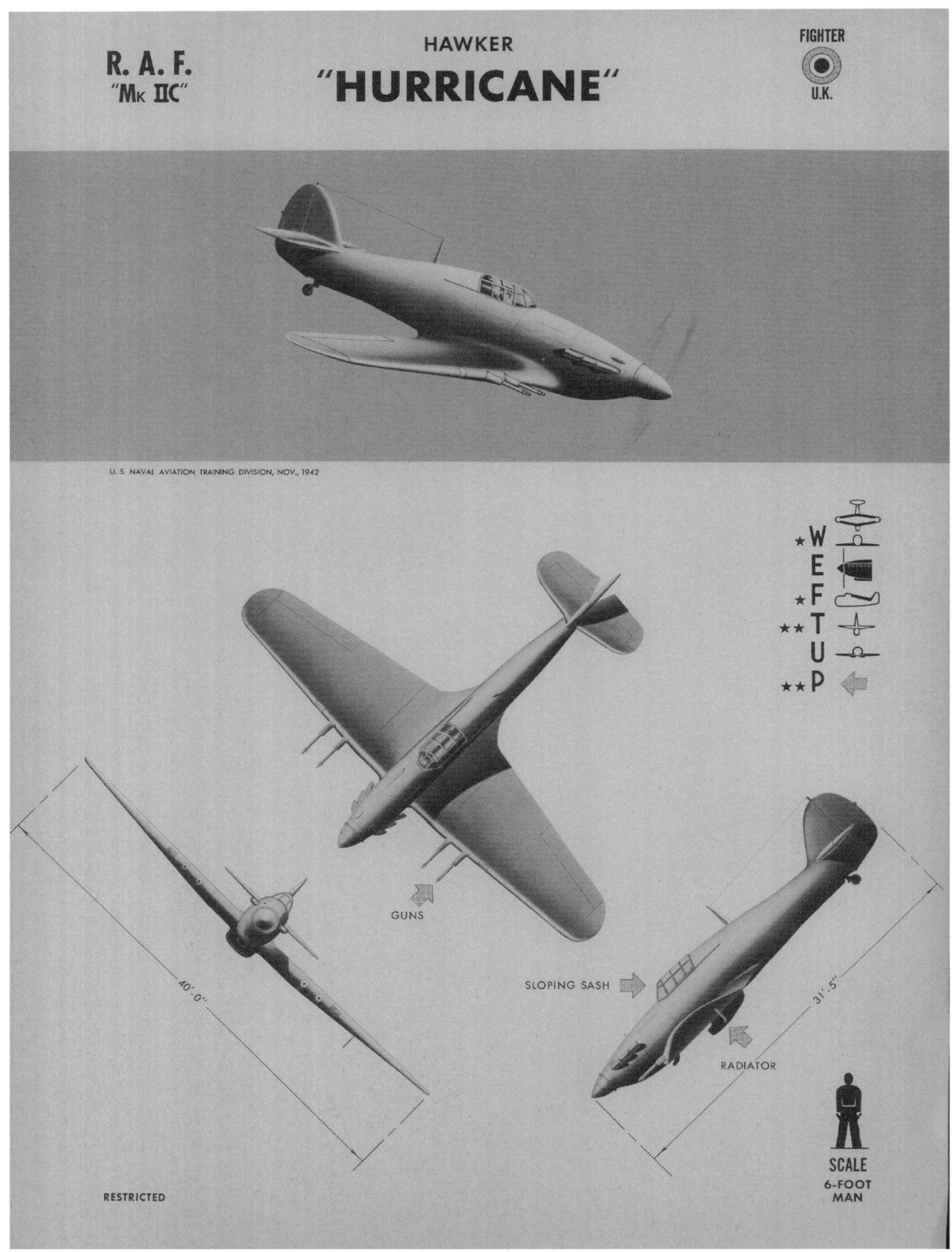

80

ODDITIES

'PIGGYBACK' HURRICANE – An attempt to piggyback a Hurricane on top of a Liberator bomber, thus allowing it to carry its own escort on longer range missions. A similar idea would have seen Hurricanes carried into battle or far off postings like gliders, being towed by cable behind Wellingtons.

'SLIPWING' HURRICANE – a biplane version of the Hurricane designed to allow the fighter to take off in a shorter distance.

SUNFLOWER SEED – A Hurricane fitted with a vertical rocket-firing device directly behind the cockpit to hit aircraft flying directly above.

THE HURRICAT

Winston Churchill famously said that after the war that the one thing which truly frightened him was the threat to Britain's Atlantic convoys. They provided an essential lifeline to an island that was very far from self-sufficient, bringing in everything from food to oil, raw materials to manufactured goods. If the Nazis could stop the convoys, they could choke Britain into submission.

With the fall of France in 1940, the Luftwaffe could now base bomber forces further south and west, enabling them to strike at shipping even further out in the Atlantic. A particular threat was posed by the 4-engine Focke-Wulf Fw-200 Condor, that boasted a range of over 2,000 miles.

Britain could provide little in the way of air cover for the inbound convoys, as the distances would be too great for virtually all their warplanes to fly. Protection from aircraft carriers sailing with the convoys was also ruled out. The Royal Navy simply did not have enough carriers.

A partial solution came in the form of the Catapult Assisted Merchantman or CAM ship, a regular convoy ship fitted with a 70 foot rocket-assisted catapult rig. A Hurricane would be set in place on a cradle and then blasted along the rig by 13 solid fuel rockets with deafening explosions and huge gouts of flame as it

✈ L-R: Hurricane ready for launch on CAM; Testing the catapult system on dry land

✈ Hurricane Mk II diagram for the US market

HAWKER HURRICANE THE HISTORY OF A LEGEND

The CAM ships would not just sail the Atlantic routes but also saw service in the Mediterranean and with the Arctic convoys to Russia. The first CAM Ship set sail as part of a convoy on 18 May 1941. Unfortunately it was torpedoed and went straight to the bottom taking its Hurricat with it. Hurricats achieved their first 'kill' on 2 August 1941, when Lieutenant Robert Everett RNVR (a former National Hunt jockey and Grand National winner) engaged and destroyed a Fw-200. More 'kills' followed, with Hurricats notching up a total of six Condors by the start of 1942 and ten during the course of the project. Hurricats also shot down two He-111s and a Ju-88. Amazingly, given the horrendous risks faced by Hurricat pilots, only one was killed during all the actions.

CAM Ships and Hurricats were gradually phased out across 1942 as more small escort carriers became available to accompany the convoys.

THE SEA HURRICANE

At the start of the Second World War, the Royal Navy found itself in desperate need of modern quality fighters. Nothing they possessed could remotely take on the Bf-109. The Hurricane had proved itself good for aircraft carrier duty the hard way – by taking off and then landing on HMS *Glorious* during the Norwegian campaign of 1940 with no certainty that it could cope. When it did, plans moved forward to properly adapt the aircraft for an ongoing carrier role. By March 1941, tests were already underway to fit an A-frame arrester hook.

The very first official Sea Hurricane – the Mk 1A 'Hurricat' – was intended solely to be used as a

rapidly accelerated to 70 mph with the hapless pilot subjected to 3.5 Gs. The cradle would then hit a set of buffers, the sudden impact throwing the fighter plane into the air where its engine would take over. The Hurricane once launched would engage any enemy aircraft. There was a small snag to the design however. There was nowhere to land the Hurricane after battle. The pilot would have to either abandon his aircraft and bail out or else try to crash land on the ocean and hope a passing ship could pick him up.

Thirty-five CAM ships would eventually be fitted with a catapult for Atlantic convoy service, and 250 Hurricanes converted for use with the catapult system. They became known colloquially as 'Hurricats'. Given the terrifying requirements of the posting, all Hurricat pilots were volunteers.

✠ Hurricane aboard a CAM ship

✕ L-R: Sea Hurricanes and Seafires on HMS *Argus*; Sea Hurricane landing on HMS *Victorious*; HMS *Glorious*

✠ 118 Squadron Sea Hurricane

SIGNIFICANT VERSIONS AND VARIANTS

catapult launched aircraft from CAM ships, but the next variant, the Mk IB came with both catapult spools and an arrestor hook so that they could be used for both carrier and catapult duties. Mk IBs' were first supplied to the carrier HMS *Furious* in January 1941, but the *Furious* did not see action with them until July 1941. When 880 Squadron of the Fleet Air Arm achieved their very first kill – a Dornier Do-18 flying boat flying reconnaissance over the North Sea That same year the carriers *Argus, Eagle, Formidable* and *Victorious* were all equipped with Sea Hurricane IBs. From October, Sea Hurricanes also began to be put aboard MAC (Merchant Aircraft Carrier) ships, which were regular cargo carrying vessels with an emergency flight deck added.

By early 1942, there were 100 Sea Hurricanes serving on board ship with 801, 802, 806, 883, 885 and 889 Squadrons. This deployment would eventually grow to 32 FAA squadrons flying the Sea Hurricane IB. A total of 340 Mk IB's were converted in the UK from regular Hurricanes by General Aircraft Ltd and a further 50 were manufactured in Canada. The Mk IIBs were best known for their service on the Mediterranean convoys during the defence of Malta. The high point of service for the Sea Hurricane IIC was Operation Torch (The invasion of North Africa) in November 1942.

By late 1943, the Sea Hurricanes were largely regarded as obsolete and were gradually replaced in service by the Martlet, the Wildcat and the Corsair. The last FAA squadron still flying the Sea Hurricane converted to Martlets during September 1944, by which time 537 Sea Hurricanes had served with the Fleet Air Arm.

HAWKER HURRICANE THE HISTORY OF A LEGEND

THE HURRICANE ABROAD

AUSTRALIA

Three squadrons of Hurricanes flew with the Royal Australian Air Force as part of the overall 'Desert Air Force' in the Mediterranean and Western Desert. Only one Hurricane ever saw service in Australia itself.

BELGIUM

Belgium originally agreed to purchase twenty Hurricanes in April 1939 on the understanding that they could be built by Avions Fairey in Belgium itself. The home manufacture did not go smoothly and the factory had only managed to turn out two fighters by the time the Germans invaded on 10 May 1940. By this time however, the Belgian Air Force had managed to obtain fifteen more Hurricanes direct from Hawkers in England. Nine of these were unfortunately caught on the ground at their airfield at Schaffen near Diest and destroyed on the first day of the German invasion. The rest perished in the subsequent fighting. After the defeat of Hitler, six ex-RAF Hurricane Mk IIBs and Cs were donated to the Belgian Air Force and served from 1946-1948.

CANADA

At the end of 1938, Canada ordered twenty Hurricanes, the first six of which arrived in Vancouver in February 1939. That same year, the Canadians agreed to commence manufacture of the aircraft at the Canadian Car and Foundry Co Ltd (Can-Car) of Fort William, Ontario. The first forty Canadian-built Hurricane Mk Is were ready by the summer of 1940 and the majority were shipped to the RAF for use in the Battle of Britain. Some served with the Royal Canadian Air Force's No 1 squadron during the battle. 340 more Hurricanes, now designated Mk X to denoted their origin, were shipped over to Britain from 1940 to 1941. A third batch comprised 149 more Mk X variants, followed in good order by 150 Mk XI models, now modified to take either twelve machine guns or else four 20mm cannons. Two final batches were supplied during the war, the first of 248 twelve-gun Hurricane Mk XII with Packard-Merlin 29 engines and the second of 150 Hurricane Mk XIIAs. The majority of this last batch were either donated to the Soviets or else served with the Allies in Burma, although some were also converted to Sea Hurricanes and served on carriers with either the Royal Navy or else the Royal

SIGNIFICANT VERSIONS AND VARIANTS

Canadian Navy. 1,451 Hurricanes would eventually be built in Canada under the all-seeing eye of female aviation heroine and Chief Engineer Elsie MacGill, the 'Queen of the Hurricanes'.

Seven Canadian squadrons flew Hurricanes on wartime operations. A further total of nine squadrons operated at home in Canada.

EGYPT
The Royal Egyptian Air Force were supplied with used Hurricanes to serve in the Allied cause. Mk Is were given over to fight with No 1 Fighter Reconnaissance Squadron.

FINLAND
In March 1940, Finland was embroiled in their intense 'Winter War' against the Soviet Union when British authorities finally permitted the sale of twelve Hurricane Mk Is to the Finnish Air Force. The war against the Soviets ended uneasily but flared up again in June 1941. The surviving Hurricanes served in the defence of Helsinki, dogfighting against Soviet-built bombers. The Hurricanes were not popular with Finnish pilots and from 1942 were phased out from frontline combat. They served as trainers until 1944.

FREE FRENCH
Elements of the Forces Aériennes Françaises Libres – the Free French Air Force – flew Hurricanes over North Africa between June 1940 and May 1943. Falling under RAF overall control, they were designated No. 341 Squadron RAF.

GERMANY
The Luftwaffe had access to a number of captured Hurricanes during World War Two and flew them to demonstrate to Luftwaffe pilots the fighter's capabilities and limitations.

GREECE
The Royal Hellenic Air Force flew two squadrons of Hurricanes while in exile in the Middle East after the fall of Greece. They were designated 335 'Tiger' and 336 'Olympus' Fighter Squadrons and fell under the auspices of the RAF. First flying Hurricane Mk Is, they served in convoy protection, bomber escort and ground attack roles and converted to Hurricane Mk

✈ L-R: RAAF Hurricane; RCAF Hurricane

✈ Royal Canadian Air Force pictured in front of a Hurricane

HAWKER HURRICANE THE HISTORY OF A LEGEND

IIBs before serving in the Battle of El Alamein. The squadrons adopted Spitfires at the end of 1943.

INDIA
The Royal Indian Air Force had eight squadrons equipped with Hurricanes during World War Two. Post war, they served on the fractious North-West Frontier until 1947, when they were withdrawn for use as trainers.

IRISH REPUBLIC
The Republic initially ordered three Mk Is to serve with the Irish Air Corps during World War Two, replacing its ageing Gloster Gladiator biplanes. A fourth followed in 1943. Later in the war, the IAC benefited from the purchase of a further seven Mk Is and six Mk IICs. The last Hurricane left IAC service in 1947, to be replaced by Seafires. Ireland unofficially possessed several more Hurricanes during the war, impounding several that had been forced to land in the Republic.

LATVIA
In 1939, Latvia ordered and paid for 30 Hurricanes, but they were never able to be delivered.

NEW ZEALAND
New Zealand flew two Hurricane Squadrons in the Pacific – 486 and 488 Squadrons. 488 Squadron Hurricanes were used for home service and eventually ended up as target decoys. 486 Hurricanes flew briefly as night fighters until being replaced by Hawker Typhoons.

PERSIA
Persia tried to order eighteen Hurricane Mk Is but only one was delivered by September 1939. Instead, the Shah had to wait until 1943 to receive any more, and then only obtained a further ten 'used' ex-RAF Hurricanes. In 1945-1946, a further sixteen Mk IICs were sent to Persia. All had their guns removed for use as trainers.

POLAND
The Poles requested a Hurricane for evaluation in the summer of 1939. They liked what they saw and ordered nine more. The German invasion in September 1939 prevented them from being delivered.

PORTUGAL
In 1943, the Portuguese government ordered fifteen Hurricane Mk IICs A further order followed in short order, so that by the end of the year the Portuguese had received an extra fifty Hurricane Mk IIBs and Mk IICs. They served with three squadrons and remained flying with the Portuguese Air Force until 1952 when they were all replaced by American P-47 Thunderbolts. That same year, some of the Portuguese Hurricanes were returned to England to feature in the feature film *Angels One Five*.

SIGNIFICANT VERSIONS AND VARIANTS

RUSSIA

A full 1/5 of all Hurricanes built were supplied to the Soviet Union. They were delivered by Arctic convoy, overland via Iran or else from stocks held in India from 1941 to 1944. Filled with gratitude, the Soviets played down their importance, moaned that they weren't getting enough of them and then complained that the aircraft was rubbish, Stalin called it 'shit'.

RUMANIA

The first order for twelve Hurricane Mk Is was received in November 1938 after King Carol of Rumania attended a display by Hurricanes in England and returned home enthused. Delivery commenced in August 1939. They were regarded as the offensive spearhead of the Rumanian Air Force, which was otherwise equipped with largely obsolescent warplanes. When Rumania joined the Axis side in November 1940, they were ordered by the Germans to gradually standardise on the Bf-109. Despite being used successfully against Soviet bombers following Operation Barbarossa, the Hurricanes were removed from front line service and converted to trainers.

SOUTH AFRICA

The second foreign nation to buy the Hurricane, South Africa initially purchased seven Mk Is over the course of 1938-1939. During the war, seven SAAF squadrons were equipped with Hurricanes, the first – No.1 Squadron – initially seeing action against the Italians in East Africa.

TURKEY

The Turkish Air Force purchased fifteen Hurricanes from Britain in 1939. There were still Hurricanes flying with the Turks post war.

✈ 488 Squadron of RNZAF
✈ Yugoslav Hurricane Squadron

YUGOSLAVIA

The Royal Yugoslavian Air Force were the first overseas customers for the Hurricane, ordering twelve in early 1938 to arrive at the end of the year. Another batch of twelve were ordered shortly after, the first of which was delivered in February 1940. The Yugoslavians were impressed and negotiated to build their own Hurricanes under licence. 100 further Hurricane Mk Is were intended to be built at plants in Belgrade and Zemun. Tests were also conducted to fit and fly a Hurricane with the same Daimler-Benz engine as a German Bf-109. By the time of the German invasion in April 1941, Hurricanes were serving with three Yugoslav squadrons. Several were caught on the ground and destroyed on the first day. Several more were lost in the confusion of a friendly fire incident when a force of Yugoslav Hurricanes met a similar force of Yugoslav Bf-109s. The remaining Hurricanes held out for a week against the Luftwaffe, after which the survivors were deliberately destroyed by their ground crews. After the war, sixteen Hurricanes served with the SFR Yugoslav Air Force.

FAR-FLUNG FIELDS

NORTH AFRICA AND THE MIDDLE EAST

The first tropicalized Hurricane – L1669 – arrived in Khartoum in the Sudan in September 1939 to be tested in a harsh desert environment. Tests went well and were declared '*very satisfactory*'.

When Italy declared war in June 1940, L1169 was still the only Hurricane in the theatre and so was flown seemingly endlessly from air strip to airstrip in Libya and Egypt to try and fool any Italian spies that Hurricanes were plentiful.

The first four Hurricane reinforcements did arrive in Alexandria that same month, but they were swiftly redirected to Malta, which was suffering sustained bombing raids by the Italian Air Force. Their place in the defence of Egypt was filled by four more Hurricane Mk Is, which were crated up and shipped into Alexandria to become part of 274 Squadron. In September, five more arrived, flying overland this time from the port city of Takordai in Ghana, ranging 4,000 miles over jungle and desert up to Egypt via numerous little airstrips along the way. Due to a clerical error, these aircraft had not been tropicalized or fitted with dust filters and had their engines damaged beyond repair. New engines were duly fitted.

The lesson had been learned by the time the next Hurricanes arrived in Africa, and by mid-September

274 Squadron could field twenty-four of the aircraft. Late-November saw a further thirty-four Hurricanes delivered to Takoradi aboard the carrier *Furious*, then flown up to Egypt bound for 73 Squadron. Over 1,500 Hurricanes would be supplied by this route before the end of 1942.

Middle East Command had by now spotted Italian forces building up in strength on the border between Libya and Egypt and an invasion seemed imminent. One of the Hurricanes from 274 Squadron was pressed into service to monitor the situation and was fitted out with cameras. It became the first Hurricane to serve as a reconnaissance aircraft and was designated Tac R Mk I. In mid-September 1940, Italian forces crossed the Egyptian border but with some timidity and their progress was slow.

It was at this time that Arthur Tedder, Deputy Commander of the RAF in the Middle East, began experimenting with strapping 40lb bombs to the wings of his own personal Hurricane. Quite accidentally he had invented the 'Hurribomber' and extended the life of the aircraft for another five years.

The future of the almost obsolescent fighter would lie in being a first rate ground attack aircraft, with the first 'official' models joining the fight in May 1942.

Both 274 and 73 Squadrons were still under strength when General Wavell decided to hit the Italians with a ferocious counter-attack codenamed 'Operation

✈ HMS *Furious*

✈ L-R: 274 Squadron pilots in Egypt; Deputy Commander Arthur Tedder

✈ RAF operations in the Middle East

Compass' in Libya on 9 December. Teaming up with the Gladiator Squadrons, those Hurricanes that were available strafed Italian airfields and road transports. One journalist reported them as flying just 30 to 40 feet above the ground on their strafing runs. They also provided powerful air cover against the outmatched Italian Air Force. The campaign culminated at the Battle of Beda Fomm, where 130,000 Italian soldiers were captured as they tried to flee from Benghazi. During the whole campaign, the RAF lost just six Hurricanes (and a further five Gladiators) while accounting for 58 Italian warplanes. It's claimed that one Hurricane pilot, Flight Lieutenant Charles Dyson, shot down seven enemy aircraft in a single attack, a feat no other RAF pilot has ever equalled. The immediate threat to Egypt was over.

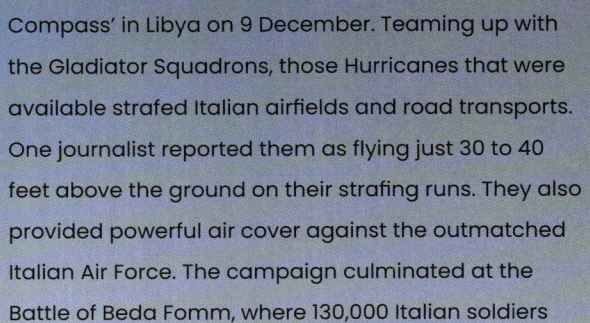

FAR-FLUNG FIELDS

The Hurricanes of three SAAF Squadrons, operating from improvised desert air strips, also played a leading role in the skies over Abyssinia, Eritrea and Somaliland as British and Commonwealth forces drove the Italians out of their East African empire bases. They proved successful in both fighter and ground attack roles. Complete victory in East Africa was achieved by the end of 1941.

'One squadron of only nine Hurricanes destroyed over 100 enemy planes.'
– The New York Herald Tribune

✈ L-R: Hurricane maintenance in North Africa; Hurricanes fly above the African desert
✈ British Generals assess the desert plains
✈ Hurricanes needed to be tropicalized to cope with the harsh desert conditions

WAR IN THE MED

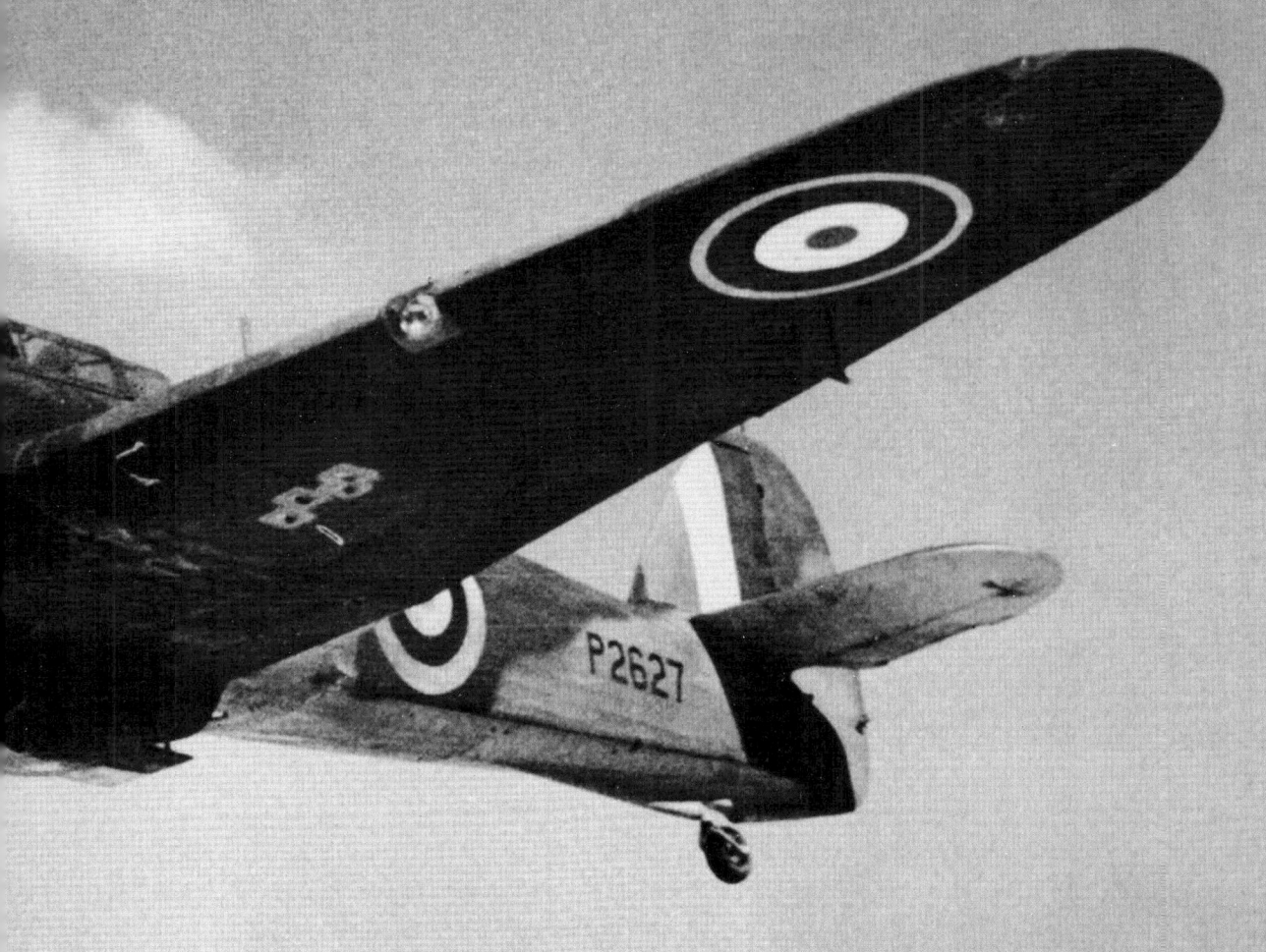

HAWKER HURRICANE THE HISTORY OF A LEGEND

MALTA AND THE CONVOYS

Italy entered World War Two on 10 June 1940 with dreams of greatly expanding its empire in Africa. To do this, it would require secure supply lines across the Mediterranean – and the tiny island of Malta stood squarely in the way. Just 50 miles south of Sicily, it posed a threat to fascist ambitions that had to be smashed. Within hours of declaring war, Italian bombers were already hammering at the island, concentrating on the British airfields there. Although the defenders could only put up three Gloster Gladiator biplanes at a time, the raid and those that followed were largely failures.

Aware of Italian ambitions against Egypt in particular, Britain rushed a small number of Hurricanes to Alexandria to strengthen defences in early June 1940, but four of these were almost immediately sent on to strengthen Malta's defensive capabilities. The island now had seven defenders, referred to as the 'Malta Fighter Flight'. The Italian Regia Aeronautica, by contrast, could field over 200 warplanes against the island. By the start of August 1940, such was the ferocity of the air war over Malta that the RAF defenders had dwindled back down to three. Twelve Italian planes had been shot down by the defenders.

As the Battle of Britain raged, twelve new Hurricanes Mk Is adapted for tropical use were spared to reinforce the defence of Malta. They were shipped out on board the carrier *Argus* as part of 'Operation Hurry.' None of the aircraft were adapted for carrier operations but it was anticipated that they could take off safely after the recent experiences of 46 Squadron off Norway. *Argus* took the Hurricane force as deep into the Mediterranean as Sardinia, before launching them in two flights of six on 2 August. All twelve made it safely to Malta where they formed the backbone of 261 Squadron. The ground crews travelled separately and were delivered to the island by submarine. The new Hurricanes helped to crack the morale of the Italian Air Force by destroying 37 raiders. All Axis daylight raids were called off.

Because of the Battle of Britain, no more Hurricanes could be spared until November 1940, when the carrier *Argus* once more took twelve Hurricanes

A Gloster Gladiator biplane; RAF Hurricanes in Malta

Benito Mussolini

Using a sextant aboard the aircraft carrier HMS *Argus*

into the Mediterranean. This time the aircraft were launched early at the very limit of their fuel supply because of the threat of Italian naval forces ahead. The operation quickly turned into a disaster. One by one the Hurricanes began to run out of fuel and drop into the sea. Only four of the twelve made it safely to Malta. The other eight ran dry and were lost. Of the eight pilots, just one was saved.

The aerial siege of Malta continued with whatever Hurricanes and Gladiators the defenders could muster. A lack of spare parts kept many of the surviving Hurricanes grounded, but 261 Squadron could claim a total of 45 Italian warplanes by the end of 1940. The Italians admitted to another 194 damaged. The losses convinced Mussolini that Britain had a much greater force of fighters on the island than was really the case, and he became

overcautious as a result. Invasion plans were shelved while the RAF took advantage of the timidity of the Italian Air Force to further reinforce Malta with whatever fighters and bombers they could muster.

THE BALKANS CAMPAIGN

In 1940, Mussolini grew rather uncomfortable at the continuing prospect of being the junior partner in his alliance with Adolf Hitler. He was noticeably late to the Battle of Britain, where his meagre air force contribution was swiftly chopped to pieces by the RAF. This did not deter him. He had already invaded Albania in 1939 and the following year would see him invading France (after Hitler had already subdued it).

As part of what Mussolini called his dream of *Spazio Vitale* (living space) for the fascist Italian people, he looked to Greece with increasingly avaricious eyes, and started to goad and provoke them. The actual invasion came on 26 October 1940. Anticipating the attack, the RAF had flown in a number of Wellington and Blenheim bombers with Gloster Gladiator biplane fighter escorts to support Greek air and ground

✈ Vickers Wellington; Hurricanes of 261 Squadron take off from RAF Ta'Kali Malta; Hurricane landing at Araxos Greece

forces. In just a fortnight, the Italians were in serious trouble. The Greeks had held them and were now in the process of an ambitious counter-attack, crossing into occupied Albania and seizing sizeable chunks of territory in the south. Only lack of supplies and increasingly foul weather held the Greeks back. They decided to wait until Spring as indeed did the Italians. When Spring came, the Italians were first off the mark, but their ambitious spring offensive crumbled in just days, with Italian forces suffering roughly ten times the casualties of the Greek defenders. It was time to ask the Germans for help.

Hitler was not at all amused by Mussolini's antics. He denounced the invasion of Greece as *'this stupid show'* and *'pointless'*. The Fuhrer was furiously building up resources for Operation Barbarossa – the invasion of Russia – and now he would have to delay the attack to pull Mussolini's fat out of the fire. The German response, when it came, would be overwhelming.

The RAF had had a moderately successful war so far in the Balkans. Their bombers had only had a limited effect but the Italian Air Force had proved so worthless in combat that the RAF's antiquated Gloster Gladiator biplanes had by the end of 1940 succeeded in shooting down 42 Italian warplanes for the loss of just six.

Hurricanes had first arrived in February 1941 and swiftly achieved total air superiority. 80 and 112 Squadron were converted and 33 Squadron newly flown in. During the Greek attack at Tepelini alone, the three squadrons of Hurricanes flying in support managed to destroy 27 Italian fighters and bombers.

HURRICANE ACES

WILLIAM 'CHERRY' VALE

Born in Kent, Bill Vale first joined the RAF in 1931 as a fitter and then moved to air crew as a gunner. By now posted to Egypt, he decided to train as a pilot and flew first Hawker Harts and then Gloster Gladiators.

After Italy declared war, he proved an effective pilot with No 33 Squadron before being transferred to 80 Squadron with whom he fought over Greece and the Balkans until April 1941. The Squadron had converted to Hurricanes at the start of the year and it was while flying Hurricanes that he claimed 20 kills in this theatre. He also won a Distinguished Flying Cross for his later actions defending Crete. As Greece and its islands fell to the Germans, Vale escaped to Egypt and saw action against the Vichy French during the invasion of Syria where he claimed three French warplanes shot down.

Vales exact tally of confirmed kills is somewhat confused due both to inept records kept by the Italians and the chaos following the invasion of Greece, but he is generally credited with 30 'kills', plus three shared.

HAWKER HURRICANE THE HISTORY OF A LEGEND

By the end of March, 93 Italian warplanes had been accounted for, and the Hurricane once more proved outstanding in a ground attack role.

When the German invasion of Greece began on 6 April 1941, the situation changed drastically. German forces some 42 divisions strong burst into Northern Greece and swept through Yugoslavia to attack Greece on a different front. It was impossible now to stop the Axis forces surging south. Suddenly the Luftwaffe were everywhere and in overwhelming numbers, just as they had been during the Battle of France. The RAF and Greek Air Force were both completely overwhelmed by a Nazi force 1,200 aircraft strong. There were just four Hurricane squadrons in theatre now, and all had received casualties.

Hurricane missions ceased to be escort and support focused, and the meagre force instead tried their hardest to provide some opposition at least to the German bomber formations filling the skies. Hurricane pilots might be called to fly five or six combat sorties every day during those desperate days of April – and their numbers were severely limited.

The author Roald Dahl, himself a Hurricane pilot stationed in Athens, recalled the sheer desperation of those last days in a vivid account of his small band of Hurricanes tackling over 200 Luftwaffe bombers

✛ Hurricane pilots relax before their next mission in the Balkans

and fighters over Athens. Having survived one more dogfight, he returned to base;

'As I walked across the grass I suddenly realised that the whole of my body and clothes were dripping with sweat. Then I found my hand was shaking so much I couldn't put the flame to the end of the cigarette. The doctor, who was standing nearby, came up and lit it for me.'

During this time 40,000 British soldiers were evacuated to the island of Crete under intensive German bombardment. The evacuation order came on 21 April with just 20 surviving Hurricanes available to prevent the Luftwaffe from turning the retreat into a massacre. The number of defenders then dropped dramatically as Bf-110s caught the squadrons on the ground and destroyed 13 of the 20 Hurricanes. The surviving seven joined the evacuation to Crete.

HURRICANE ACES

SQUADRON LEADER MARMADUKE THOMAS ST. JOHN 'PAT' PATTLE

'A G.50 came for us and in a flash a Hurricane just shot it off our wingtip. It was wizard.'
– Blenheim gunner recalling the moment 'Pat' Pattle saved his bomber, 20 February 1941

'I never had a chance – we came up right behind this pair of 109s – Pat gave them a left and a right and it was all over. Both went down in flames.'
– Pattle's wingman recalls his lightning reflexes, 6 April 1941

A diminutive and quietly spoken man, South African Marmaduke Pattle was the RAF's highest scoring ace in the air war over Greece.

Originally rejected for air service by the South African Air Force because of his lack of flying experience, he emigrated to Britain as his own expense and joined the RAF instead. Pattle entered the Greek theatre as part of 80 Squadron flying an antiquated Gloster Gladiator biplane, but still managed to become an air ace shooting down five Italian aircraft between 2-4 December 1940. In February 1941, his squadron converted to Hurricane Mk Is and a month later Pattle – now commanding three Hurricane Squadrons – was awarded a DSO and promoted to Squadron Leader.

When the Luftwaffe turned up in overwhelming numbers from April 1941, Pattle's squadrons were forced back to airfields in Athens. Within scant weeks Hurricane losses mounted quite severely fighting against the superior Bf-109E. Despite being taken sick with fever, Pattle refused to stop flying, instead engaging the Luftwaffe in sortie after sortie. He became exhausted, Flying Officer Roald Dahl observed;

'He possessed the deeply wrinkled doleful face of a cat that knew all nine of its lives had already been used up.'

Despite his condition, on 19 April he was officially credited with no less than six 'kills'. On the day of 20 April 1941, Allied ships were attempting to evacuate British forces from Piraeus Harbour and every Hurricane was needed in the air to fend off the German bombers and their fighter escorts. Pattle insisted on leading mission after mission against the overwhelming Luftwaffe fleets, despite medical orders grounding him with fully blown influenza. Pilots recalled seeing him huddled on a couch in the mess, shivering under blankets and he had had to be physically helped into his flying gear. By the end of the 20 April, five out of 12 surviving Hurricanes had been shot down. Pattle's was amongst them. His Hurricane was last seen plunging into the sea at Piraeus Harbour after tangling with a Bf-110.

Marmaduke Pattle only spent nine months in active combat, but during that time managed to achieve

✈ German Bf-110

an official tally of 34 'kills', the vast majority flying Hurricanes. In the chaotic air war over Greece, he may have scored as many as 61 kills. All official records were destroyed as Greece fell to the Germans. If he did achieve 61 kills however, that would have made him by a long stretch the highest scoring RAF fighter ace of the war. As it stands, his official tally made him the highest scoring Hurricane pilot of the war.

As Greece fell, the Luftwaffe was unleashed against Crete at the start of May. They flew in 430 bombers and 180 fighter aircraft. All the RAF could put up against them were the seven surviving Hurricanes from Athens, plus nine more hurriedly flown in from Egypt and a smattering of Blenheims, Sea Gladiators and Fleet Air Arm Fulmars. For Edward Howell, his first experience of flying a Hurricane came when he jumped into the cockpit of one while under attack from five Bf-109s strafing the airfield at Maleme on Crete. Astonishingly, he managed to shoot down one of the Bf-109s while simultaneously trying to familiarise himself with the Hurricane's cockpit and controls. Finally, the order was given to evacuate every aircraft that remained – four Hurricanes and two Sea Gladiators – and get them off the island.

The Nazi paratroopers came three days later on 20 May. British forces were evacuated from the island beginning on 27 May and those that could not escape surrendered on 1 June. By the time Crete fell, 28 Hurricanes had been lost in its defence.

HAWKER HURRICANE THE HISTORY OF A LEGEND

MALTA AT BAY

By the start of 1941, Hitler had also grown exasperated by Mussolini's calamitous handling of the war in North Africa, and sent significant German forces in to rectify the Axis' losses. The Afrika Korps were despatched to Africa in February 1941, supplying them adequately would be the key to future victory. This still left Malta squarely in the way.

To counter the threat from British forces in Malta, the Germans despatched Fliegerkorps X (Flying Corps Ten) to Sicily to attack Malta. By 10 January 1941, there were 255 German fighters and bombers poised to strike, with Ju-87 Stuka dive bombers taking the lead.

Malta now could only put up sixteen Hurricanes and two Gladiators in its defence. The RAF desperately tried to reinforce 262 Squadron with whatever planes they could muster – but they were pitifully few. Just six arrived in January. Twelve further Hurricanes arrived on Malta in March 1941, delivered by the *Ark Royal*. However, five of them were lost in combat that same month. The increasingly battered Hurricanes, kept flying only by cannibalising damaged planes, found the new German Bf-109E-7 fighter a real challenge. The Hurricane was clearly inferior to its German challenger and losses swiftly mounted.

More Hurricanes were brought in by *Ark Royal* during April, at first a flight of twelve and then twenty more

later in the month, all fitted with 40 gallon long range fuel tanks now to prevent a repeat of the dreadful debacle of the previous November. Despite the reinforcements, the Germans maintained air superiority. By May 1941, much of Malta's civilian population had been evacuated out of the cities. Such was the strength of the Luftwaffe that vast swathes of the Mediterranean became effective no-go areas for British shipping.

Malta seemed poised to collapse. In the event, it was only saved by more of Mussolini's tactical blundering. His invasion of Greece had gone so badly that Hitler was forced to pull Luftwaffe air power away from the Malta campaign to assist Italy with the war in Greece. The respite allowed Malta's new Air Officer Commanding, Air Commodore Sir Hugh Lloyd, to take stock of the situation. He needed to both defend the island and strike out at Axis sea lanes

✈ HMS *Ark Royal*

✈ L-R: Afrika Korps convoy; Korps surveillance; Junkers Ju-87

✈ Afrika Korps Panzer

✈ Sir Hugh Lloyd

as best he could with whatever he could muster. German land forces were making significant gains in Africa at the time and disrupting their supplies was of critical importance. He needed numbers.

In May 1941, no fewer than 47 Hurricanes were flown into Malta as part of Operation Rocket and the *Ark Royal* despatched more Hurricanes in June. The Hurricanes were it seemed arriving with some regularity now – and the Mk II had very largely replaced the Mk I. The island could now boast three Hurricane squadrons – 185 (formerly 261), 126 and 249, all flying Hurricane Mk IIBs. However, disquiet was growing on the island about their efficacy. As early as May 1941, Sir Wilbraham Ford, Vice-Admiral of Malta, was complaining that the heavy losses suffered by the RAF that year were *'the result of inferior aircraft to those of the enemy'*. By 'inferior aircraft' he meant the Hurricane Mk I and he pulled no punches;

'Unless modern fighters are exchanged for the Hurricane Mk I and sent immediately we are riding for bad falls and will be unable to protect harbour shipping or population.'

Churchill chose to ignore him, defending the Hurricane as best he could. In the background, RAF High Command had fears that if Spitfires were sent, they would not stand up well to the desert conditions. In truth, those responsible for Home Defence were reticent to spare any Spitfires to go overseas. They would however spare Hurricanes, which could

✝ Hurricane night fighters at Ta'Kali Malta; Vice-Admiral of Malta, Sir Wilbraham Ford

be produced faster and cheaper and were now approaching obsolescence anyway. The Hurricane was lauded as more robust, and Malta would have to keep calm and carry on like a good chap.

A major convoy initiative in July 1941, aptly codenamed Operation Substance brought 65,000 tonnes of supplies to Malta that month, including vital spare parts for the Hurricanes as well as more new aircraft. With the skies now swarming with German aircraft, the approximately 450 mile flight off the carriers to Malta was now even more dangerous. All the aircraft had to fly with no ammunition, as the Hurricane cockpit was so small that no belongings could be stowed there for the flight. Instead, some few personal items like shaving kits and soap were stored in the wings instead of ammunition containers. Tom Neil of 249 Squadron recalled;

'If the enemy was sighted en route, we would be able to give them a quick squirt of Macleans toothpaste from all eight guns!'

By the end of the month, Malta had an offensive bomber force of some 60 aircraft, supported by a Hurricane force 120 strong. With the Germans now largely absent having finished with Greece and then moving swiftly on to assault Russia, the RAF went on the offensive with everything it had to support Royal Navy warships and submarines. The second half of 1941 saw an estimated 60% of Axis shipping in the area sent to the bottom. In Africa, Axis forces started to become desperately short of supplies. The RAF also felt confident enough – with the Germans gone – to go on the occasional offensive and actually attack Italian targets on Sicily and North Africa, including road and rail transports and airfields. By the end of 1941, the Royal Navy virtually had the ports of Libya under an effective blockade, with 63% of all Afrika Korps supplies destroyed in November.

Then a series of setbacks (which inclued the sinking of the carrier *Ark Royal* by a U-boat) threatened Britain's hard-won naval dominance once more. To add to the crisis, the Luftwaffe once more turned its attention to Malta in December 1941. They arrived in strength, reinforced by aircraft from the Eastern Front. The German High Command incorrectly anticipated that total Russian collapse was imminent. They brought with them Ju-88 and Bf-110 night fighters and, in just a matter of weeks, the Allied bomber force on Malta was torn to shreds.

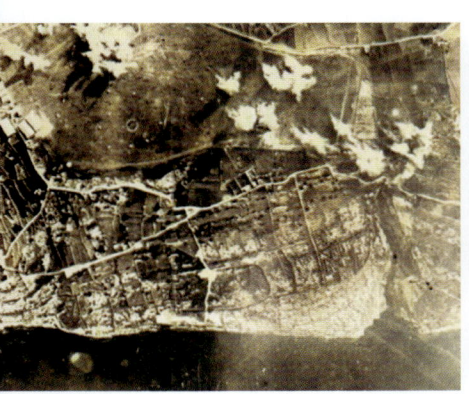  

✠ L-R: Hal Far in Malta under attack; German Junkers Ju-88; German troops carry large bombs that had been adapted for the Bf-109

FRIENDS AND ENEMIES

COUP

At the end of April 1941, Iraq suffered a pro-Nazi coup. Regional Allied bases in Iraq had to be defended by trainer aircraft adapted to carry bombs until Gladiators and then Hurricanes could reach the area. The four Hurricanes, newly arrived at 94 Squadron, were based now at Habbaniya, which became the target for Luftwaffe bomber raids by He-111s supported by Bf-110s. The Germans were operating out of Mosul, which was beyond the range of the Hurricanes and it wasn't until two Hurricanes with long range drop tanks were sent as reinforcements that the RAF could strike back. When they did, the two Hurricanes achieved total surprise and destroyed a number of enemy aircraft on the ground both in Mosul and Erbil. One of the Hurricanes was shot down however with the loss of its pilot as it launched a strafing run. Four more Hurricanes arrived in short order to reinforce Habbaniya, along with the specially adapted Hurricane reconnaissance version Tac R Mk I.

The coup collapsed in just one month.

SYRIA

'...the best time to attack was lunchtime because the French being the French always had their break then.'
– Ben Bowring, 80 Hurricane Squadron.

The French Vichy government in Syria had been caught collaborating with the Nazis during the fighting in Iraq and plans were swiftly drawn up to invade both Syria and the Lebanon. Hurricanes started to appear on the air base at Haifa and on Cyprus to provide support. They skirmished with French Vichy aircraft as tensions rose. As Allied forces launched their invasion on 8 June 1941, Sea Hurricanes of the Fleet Air Arm 806 Squadron. provided air cover above Haifa while the RAF 80 Squadron Hurricanes attacked the French air base at Rayak.

The next day, 80 Squadron Hurricanes from Cyprus flew into Haifa and joined the sporadic air battles with Vichy French aircraft. The fortunes of the invasion went back and forth and instead of crumbling as the French were expected to do – the next few days saw

L-R: Arming a Hurricane Tac R; Hurricane diving off for its next strike

the Vichy forces flying in more warplanes. The Allies responded by sending in Tomahawks to reinforce their Hurricanes. Aircraft losses mounted on both sides. More Hurricanes, this time out of Abu Sueir, turned up on 20 May. They joined other fighters in theatre in strafing retreating French troop columns around Beirut, train transports and then the Vichy airfields at Talia and Rayak.

Four Hurricanes and a further four Gladiators were formed up as 127 Squadron to support Commonwealth forces advancing on Aleppo, but all the Hurricanes were swiftly lost dogfighting with French D.520s flying bomber escort duties. The end of the campaign came on 18 July 1941 when the French

✝ German Bf-109 maintenance

just lost heart and flew their surviving aircraft off to other safe Vichy airfields around the Mediterranean. They had lost 179 warplanes during the conflict, the majority caught and destroyed on the ground.

THE KORPS IN THE WESTERN DESERT

When the German Afrika Korps under Rommel arrived in Libya in March 1941, the whole equation changed in the Western Desert. The Luftwaffe contingent brought with them the new BF-109F and it soon proved capable of easily outclassing the Hurricane Mk I in a dogfight. When the Mk II started to replace the Mk I in May and June of 1941, the aircraft still struggled mightily to compete. The Mk IIs were immediately designated to do the lion's share of the fighting, while the Mk Is were given over to safer tasks.

Earlier than expected. Rommel attacked, surging forward in late march, In just a few weeks, British and Commonwealth forces were driven out of Libya all the way back into Egypt as the Germans swept east. The only Hurricanes in theatre to oppose them were No 73 Squadron and 3 RAAF. They were mostly occupied with tackling the Ju-87 Stuka dive bombers providing close air support to the advancing Afrika Korps. 73 Squadron concentrated on defending the port city of Tobruk – and enjoying considerable success. A further force of reconnaissance Hurricanes from No 6 Squadron flew on mainly artillery support duties, providing target information. Both were swiftly ground down by the relentless German assault. The arrival of Bf-109F fighters effectively finished off the last Hurricane defenders of Tobruk. All the surviving fighters and reconnaissance aircraft had been withdrawn by 10 May, as the advancing Germans turned their attention from invading Egypt and instead concentrated on the besieged Allied forces trapped inside Tobruk.

THE SOVIET UNION

When Germany began its invasion of the Soviet Union in June 1941, Winston Churchill immediately contacted Josef Stalin to ask what sort of war aid he could provide. Stalin said that he wanted thirty of the finest divisions of the British Army and 3,000 fighter planes. When he had presumably stopped laughing, Churchill told him he could have some Hawker Hurricanes instead. In negotiations, two hundred was the figure initially agreed on.

Good to his word, Churchill had two convoys ready to steam north to Murmansk up in the Arctic Circle

✈ Soviet leader, Josef Stalin
✈ 151 Squadron unpacking a crated Hurricane
✈ Erwin Rommel
✈ Hurricane of 134 Squadron at Vaenga airstrip

by mid-August 1941. The first convoy transported over 500 RAF ground and air crew, along with sixteen Hurricane IIBs in crates on board merchant ships. The second convoy was formed around the navy carrier *Argus* with 24 Hurricane IIBs ready to fly off its deck. Both arrived safely.

The crated Hurricanes were all assembled in less than two weeks, while the contingent on board the *Argus* took off and flew to their new base at Vaenga. RAF pilots then gave a demonstration of the aircraft's capabilities to a mass gathering of the Soviet Union's great and good, with low level flying that had grand admirals and top army brass scuttling for cover and hugging the ground.

However impressive the display, the RAF Hurricanes experienced problems from the start. The state of

the airstrip at Vaenga was pitiful, tools and spares mysteriously went missing, guns jammed due to their grease solidifying in the freezing climate and the Soviets supplied the Hurricanes with the wrong octane fuel, causing two to have potentially serious engine malfunctions in mid-flight. An ongoing battle with dysentery led the British contingent to vengefully name their latrine 'The Kremlin'. Regular temperatures of − 26 degrees C did nothing to improve the situation, especially if someone had to *'pop out to the Kremlin'*.

By 11 September 1941, Hurricanes flown by RAF pilots were on patrol over Murmansk, protecting both the port and the Murmansk-Moscow Railway. The very next day they found themselves dogfighting with Bf-109s. Thoroughly surprised by the presence of the Hurricanes, four Bf-109s were swiftly shot down for the loss of one Hurricane (flown by Sergeant Pilot N. H. 'Nudger' Smith, aged 18).

When not on defensive duties or giving Russian pilots flying lessons, the RAF Hurricanes flew as escorts for Soviet bomber formations – but already had trouble keeping up with them. Despite the Hurricanes accounting for sixteen enemy aircraft during this time, age was beginning to tell. As the weather worsened, flying dropped off drastically. Bored pilots and ground crew took to mucking about on makeshift sleighs – until 155 Wing suffered so many accidental injuries that the practice was forbidden.

By 22 October, all the Hurricanes in 155 Wing were formally given over to Soviet pilots of the Naval Air Arm.

CRUSADER

Churchill was still feeling there was opportunity aplenty in the Western Desert – if only it could be seized. Commonwealth troops moved forward on 18 November 1942 with the aim of relieving the besieged garrison at Tobruk and then heading west towards Tripoli. The Allies enjoyed a huge advantage in armour, while Rommel was desperately stretched due to Malta blocking his supplies,

RAF forces available to support the campaign – now called 'Operation Crusader' – stood at 29 Squadrons based forward in the Western Desert region plus a further eleven dedicated to the defence of the Suez Canal. 25 flew Hurricanes.

L-R: Hurricanes of 151 Squadron in the Russian skies; Operations in the Soviet Union; Conditions were difficult for the RAF pilots

FRIENDS AND ENEMIES

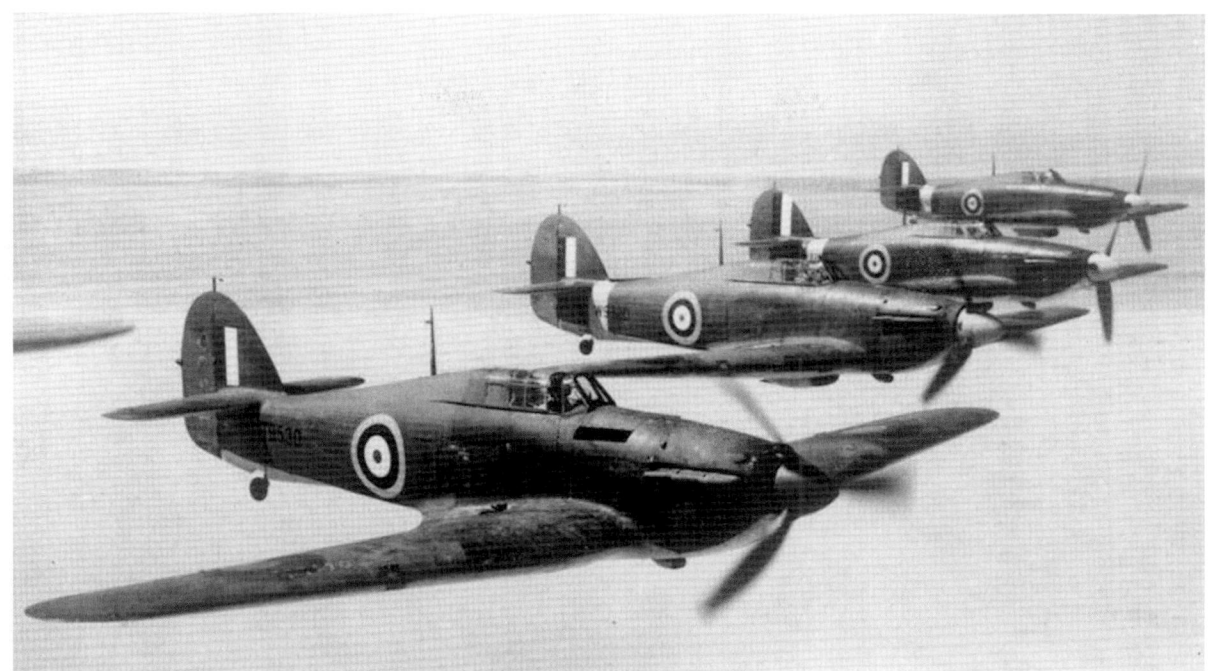

Crusader began with a series of intense air attacks on Axis forces inside Libya by squadrons of Hurricanes and Tomahawks. Tobruk was relieved in December and by January 1942 Hurricane Mk II B's and C's began to appear in theatre. It was the advent of the Hurribomber in theatre, packing 40 or 250lb bombs. All that held them back was telling friend from foe both of whom wore desert camouflage. In the air, Hurricanes in their traditional role decimated Stuka formations.

The 8th Army fought on with the intent of seizing Benghazi – but were held at El Aghella. Meanwhile, the Luftwaffe's renewed attacks on Malta allowed Rommel to gorge on fresh supplies. The tide turned. Again.

✠ ABOVE: Hurricane formation over the desert, ready for action L-R: Preparing to take off in a trail of dust and smoke; Hurricanes in North Africa with added camouflage bite

1942

HURRICANES IN THE FAR EAST

When the Japanese attacked the British in Malaya on 6 December 1941 and the Americans at Pearl Harbor the day after, it came as a rude wake up call. British possessions in the Far East were being defended by relatively scant resources. When aircraft got to become obsolete, they were despatched off out east where the threat was perceived to be very much less. Malaya and Singapore relied for their defence on such obscure types as the Vickers Vildebeest and the Brewster Buffalo (a fighter despite having only two machine guns), along with aircraft that had proved disastrous when ranged against the Luftwaffe, like the Bristol Blenheim but were still expected to be more than a match for any Far Eastern power.

That attitude of shallow overconfidence persisted even as 51 old Hurricane Mk I and IIAs were hastily diverted away from their original destination – Iraq –

✈ ABOVE: Vickers Vildebeest target practice L-R: Brewster Buffalo; Japanese Mitsubishi Zero fighter

and rerouted off to deal with the new threat. No one expected the Japanese to be able to manage them. Sir Paul Maltby, then the commander of Singapore's air defence recalled, *'It was confidently expected that the Hurricanes would sweep the Japanese from the sky.'* Churchill too thought the 51 Hurricanes would decide the outcome of the conflict. The Hurricanes arrived in Singapore and were hastily unpacked and assembled on 19 January 1942, forming 232 Squadron and 488 New Zealand Squadron. The very next day they were in combat over Singapore. They caught the Japanese completely by surprise and shot down eight unescorted bombers for no loss. The Japanese would not make the same mistake again.

The next day the bombers returned, but this time with Mitsubishi Zero fighters as their escorts. The Zeros exacted a humiliating defeat on the Hurricanes, destroying five without loss. The RAF, it seems had underestimated both the quality of the Japanese fighters and their pilots. The Hurricanes in the western theatre had at least been able to out-turn the Bf-109s. Here in the Far East, it proved that the Zero could out-turn the Hurricane. The Zeros were also proving to be much faster than the Hurricanes, which were slowed up by the tropicalized air filters. Desperate ground crews stripped two guns off of each side of the Hurricanes' wings and took out a petrol tank to try and improve the Hurricane's speed and agility, but it was too little too late.

By the end of January, of the 51 Hurricanes delivered, only 20 remained – and of these just ten were serviceable and able to fly and fight. As Japanese road convoys advanced on Singapore, the Hurricanes launched desperate ground attacks on them, but their numbers were far too few to have much effect. By 10 February, just seven Hurricanes were left and the fall of Singapore was imminent. The surviving Hurricanes were evacuated to Sumatra, itself under Japanese siege.

Sumatra was already defended by 48 Hurricane Mk IIAs, delivered from Aden by the carrier *Illustrious*, but they were experiencing the same difficulties in combat as the Singapore contingent. Worse still, 30 of them had been caught on the ground and destroyed in a series of Japanese bombing raids. The main airstrip at Palembang was knocked out and then seized by Japanese paratroopers forcing the remaining Hurricanes to congregate on a secret back up jungle airstrip designated P.II further south. The

✝ L-R: Rocket preparations for the Hurricane; Bomb loading in the Far East; Far Eastern operations

HMS *Illustrious*

survivors fought back as the Japanese invasion force closed in. On 14 February 1942, Hurricanes together with any surviving RAF or RAAF bombers sank six Japanese troop transport ships heading for the island – but seven aircraft were shot down making the attack. The next day, anything that could fly was put into the air once again to attack more Japanese landing ships. The attack was an outstanding success, killing thousands of Japanese troops. At the same time, Hurricanes attacked a Japanese airstrip on Banka Island catching and destroying several Zeros on the ground with devastating strafing runs.

Another 39 crated Hurricanes had originally been shipped to the nearby island of Java and it was to Java that all the surviving Hurricanes on Sumatra retreated when the island fell. Here they joined Dutch pilots also flying Hurricanes, living in grass huts on an improvised airstrip. The Japanese located it and launched a major attack, both bombing and strafing the airstrip just as a number of Hurricanes were refuelling. A petrol bowser blew up, destroying six Hurricanes. When the Japanese came for Java, there was nowhere left to run. Valiant strikes against Japanese forces as they landed inflicted heavy losses and Hurricanes fought on to provide cover for Allied army units but there was no hope of turning the tide. The remaining two Hurricanes were torched on their airstrip by their ground crews on 8 March shortly before they were overrun. A few New Zealander Hurricane pilots managed to elude the invaders and escaped to Australia by ship. The rest of the air and ground crews were captured by the Japanese. Half of them would not survive Japanese captivity.

1942

The next major fight for the Hurricanes now arriving in the Far East came in early April, when Japanese warplanes flying off no less than three carriers made massed attacks against targets in Ceylon.

They were met by RAF Hurricane Mk Is and Mk IIBs now based on the island. Some had flown over all the way from Karachi or Dum-Dum. Others – some fifty Mk IIs – had arrived on board the carrier *Indomitable*. Their mission was to defend both the capital, Colombo, and harbour facilities. Once more, the Hurricanes fared little better than they had in the past. On 5 April – Easter Sunday – 53 Nakajima B5N torpedo bombers and 38 Aichi D3A dive bombers attacked Colombo, flying with 36 A6M Zeros as escorts. 35 Hurricanes rose to meet them, joined by six Fairey Fulmars of the Fleet Air Arm. They targeted the bombers over the sea before they could reach inland and tried to evade the fighters, but the escorting Zeros threw up a highly effective defensive barrier. No less than 21 of the Hurricanes were shot down, together with four of the six FAA Fulmars. Japanese losses amounted to one Zero and six dive bombers, most destroyed by Colombo's anti-aircraft batteries.

The next day, the Japanese returned to attack port facilities and an airstrip near Trincomalee with 91 torpedo and dive bombers and a 41 Zero escort. Only

✈ ABOVE: Fairey Fulmar L-R: Japanese fighter pilots; Aichi D3A bombers; Nakajima B5N

HAWKER HURRICANE THE HISTORY OF A LEGEND

16 Hurricanes were able to be sent against them – and half of them were shot down. More were lost on 9 April as they desperately tried to provide air cover for the stricken carrier HMS *Hermes* as she was swarmed by Japanese torpedo bombers. They fought in vain. The carrier was lost. Ceylon was only spared further incursions when the Japanese withdrew to regroup at Midway.

BURMA

When the Japanese came for Burma, there were just sixteen RAF Brewster Buffalo fighters and twenty-one P-40s flown by American 'FlyingTigers' volunteers (who were paid by the 'kill') to protect a country 262,000 square miles in size. Once again, as with Singapore, the solution was judged to be Hurricanes.

Middle East Command agreed to release twenty of their Hurricane Mk Is for immediate despatch to defend Burma in mid-January 1942. When they arrived at Mingaladon air base near Rangoon, such was the urgency of the situation with formations of Japanese bombers raiding seemingly at will, that the Hurricanes were immediately committed to battle with no time even for ground crews to remove their long range fuel tanks. Every single Hurricane received battle damage in the hours that followed. In the event, they could not stem the tide. The Japanese air power ranged against them – some 500 planes – was simply overwhelming and they were soon assigned the task of giving what air cover they could to the Allied troops now streaming westward in retreat.

To prevent the Hurricanes being wiped out on the ground in a single devastating air raid, the aircraft were scattered across numerous small air strips located in the paddy fields around Rangoon – but this made formatting for any decent-scale action very difficult. At the start of March 1942, the Hurricanes were assigned to launch a ground attack mission against a large force of Japanese fighters and bombers sitting on a captured air base in Burma close to the Thai Border. Of the eight surviving Hurricanes only two were able to co-ordinate for the attack. They found the Japanese aircraft conveniently all lined up and in the process of being re-armed

L-R: HMS *Hermes*; Curtiss P-40 Warhawks USAF

↑ Hurricane Mk II in flight

and refuelled. Two more – both Oscar fighters – were in the process of landing. The senior Hurricane pilot, Wing Commander Frank Carey, went for them. He hit the first with his machine guns as it crashed, so the second Oscar panicked and slammed into the ground as well. Carey was then bounced by at least five enemy fighters. He managed to get one of them before the others were upon him. In their eagerness to kill Carey however, they only succeeded in getting in each other's way and the Wing Commander was able to break off. His Japanese pursuers raced after him, chasing him for some distance above the Jungle canopy before eventually breaking off. The second Hurricane on the raid, Pilot Officer Underwood, was not so fortunate. He was hit by anti-aircraft fire over the air base and had to bail out. He spent the rest of the war as a POW.

Within days, the defence of Rangoon simply became impossible and what surviving Allied aircraft there were redeployed to Magwe civilian airport and Akyab Island. The Victorious Japanese almost immediately seized their previous air base at Mingaladon and crammed it full of warplanes. On 20 March, a force of ten Hurricanes accompanied by nine Blenheim light bombers raided the air field. They achieved complete surprise, catching the Japanese air force on the ground and destroying sixteen enemy aircraft. Retaliation was not slow in coming. By the next day, the Japanese Air Force packed 230 aircraft into the air and struck back at the Hurricane base at Magwe. When the bombing ceased, there were just eleven Hurricanes left together with six Blenheims. A week later, the Japanese turned their full fury on the other air field on Akyab Island, bombing it for three full days. Seven Hurricanes were caught up in the onslaught and destroyed.

Now, the surviving Hurricanes were all evacuated west to India, reassembling in Calcutta or else to see service in Ceylon. It was widely anticipated that, when the Japanese had finished with Burma, they would come for India next. There were small scale sporadic raids made against Calcutta by night and the Hurricane forces based around the city provided air cover in response. There was no radar in the city to assist. Mostly, they waited. Meanwhile, 50 fresh Hurricanes were being shipped to India each and every week.

When Field Marshal Wavell, Commander-in-Chief of the Indian Army, launched his offensive in occupied North West Burma at the end of 1942, he could count on ten Hurricane Squadrons in theatre to provide support. Crucially, the Hurricane was now being given a ground attack role rather than purely air defence or escort duty and supported the advance by launching strafing runs against numerous different Japanese targets ranging from troop concentrations to air bases, river barges and Sampans and supply convoys to bridges and railway lines. Their effect was such that the Japanese soon only dared to move by night.

MALTA – LIFE OR DEATH

The Luftwaffe had returned to Malta. January 1942 saw 263 bombing raids on Malta – an unprecedented onslaught. The airfields were prime targets. In that month alone 50 Hurricane defenders were caught and destroyed on the ground. A further eight were lost in dogfights. Malta's effective aircraft tally fell once more to just 28 fighters. By March, they were ranged against a Luftwaffe presence 425 fighters and bombers strong.

The Hurricane was now clearly unable to square up to the latest Bf-109Fs. It was said that German pilots would deliberately fly in front of Hurricanes and waggle their wings before disengaging, just to taunt Hurricane pilots. *'Not good for the Hurricane boys here,'* complained a pilot from 605 Squadron.

At first Air Vice-Marshal Sir Hugh Lloyd towed the party line, telling Tom Neil of 249 Squadron, *'You know Neil. It isn't the aircraft. It's the men'*

Neil recalled thinking, *''I'm going to hit this bugger!'*

Squadron Leader Stan Turner, CO of 249 Squadron was equally blunt with Lloyd;

'Either, sir, we get the Spitfires here within days, not weeks, or we're done. That's it.'

The pilots fumed. Things were so bad that even the new Italian fighter – the C.202 – was outflying their Hurricanes.

✈ L-R: Hurricane Mk II engineering near Burma; Italian C.202 fighter; Hurricane flies over India

1942

Lloyd looked on helplessly. German raids were coming in so thick and fast that the few Hurricanes could not be refuelled and rearmed quickly enough to meet the next threat. He grew to acknowledge what his pilots had been telling him.

'(The Hurricane) lacked speed and firepower. It was clean out of its class. We wanted more speed, more aeroplanes and heavier armament as the task was to kill.'

The message finally got through to the Allied commanders too. Hurricanes needed to be replaced with Spitfires – and very quickly. Tropicalized Spitfire Mk Vbs were finally landing on Malta by 7 March 1942, greeted by a Maltese population running through the streets yelling, *'Spitfire! Spitfire!'*. It was the first time that Spitfires had operated outside of the British Isles. And now they kept coming.

Meanwhile, Hitler and Mussolini drew up plans to invade Malta by sea and air – and stepped up their aerial offensive once more. Allied air bases on Malta were blitzed continually for the first three months of 1942 by as many as 300 bombers at a time. The aerodromes at Hal Far and Ta'kali were carpet bombed and almost knocked out (the latter becoming known as *'the most bombed airfield on earth'*). This forced the RAF to over-rely on Luqa. Luqa then became the Axis's number one priority. In April, the American aircraft carrier *USS Wasp* delivered 47 more Spitfires to Malta but – after just one day on the island, only eighteen of the newly delivered Spitfires could still fly and fight. The next month, the *Wasp* and *HMS Eagle* flew off 64 more Spitfires. They were again in battle within 24 hours, shooting down fifteen attackers alongside the island's remaining Hurricanes. Although no-one knew it at the time, it was a turning point.

As more and more Spitfires joined the fray, they freed up the Hurricanes to go against German and Italian bomber fleets while they concentrated on their fighter escorts – just as they had during the Battle of Britain. With five full squadrons of Spitfires now operating from Malta's air fields, Axis losses mounted – just at a time when German commanders in North Africa and Russia were both demanding many more aircraft in theatre. On 10 May alone, 65 Axis aircraft were shot

✈ L-R: Hurricanes set for action in Malta; Spitfires were heralded when they arrived in Malta

HAWKER HURRICANE THE HISTORY OF A LEGEND

down over the island. By the end of the month, the Luftwaffe could put just 83 warplanes into the field. The RAF had won back control of the skies over Malta.

The last Hurricane on the island was retired in July 1942.

CONVOY ESCORTS

Summer 1942 also saw much reduced attacks by the Axis air forces ranged against Malta. Instead they targeted the Allied Mediterranean convoys, hoping to choke off supplies to the island once more. While Spitfires may have effectively taken over the defence of Malta, Sea Hurricanes from the Fleet Air Arm still gave vital air support to the Mediterranean convoys as they ran an Axis gauntlet.

Sea Hurricanes now stationed aboard the carrier *Eagle* battled fiercely against overwhelming Axis air power during Operation Harpoon in mid-June 1942. Joined by Fairey Fulmars from the carrier *Argus*, they fought against 150 Axis dive bombers escorted by a swarm of enemy fighters 100 strong. The convoy was badly chewed up, but enough ships still made it to Malta to deliver 15,000 tons of ammunition and food supplies.

The battle was even more fierce in August 1942 during Operation Pedestal when a huge convoy escorted by three aircraft carriers tried to break through again to Malta. 39 Sea Hurricanes were set against more than 500 axis fighters and bombers in an epic aerial engagement that raged for four days. The Sea Hurricanes acquitted themselves heroically, shooting down as many as 40 Axis aircraft for the loss of just four of their own. Once more the convoy suffered heavy damage – losses included nine merchant ships and an aircraft carrier – but again many thousands of tons of supplies together with the contents of an oil tanker were successfully delivered to the island.

One final prolonged air assault on Malta during October 1942 convinced the Luftwaffe High Command that Malta was now just too well defended. It cost them 34 Ju-88s and 12 Bf-109s to find out the hard way. The siege of Malta was now effectively over.

THE WESTERN DESERT WAR

After Operation Crusader, Rommel's devastating counter attack followed in January 1942, launching the Afrika Korps' armoured columns eastward again – almost trapping six Allied squadrons on their forward air bases. Four of the squadrons were comprised of Hurricanes. If they succeeded, it would have meant the decimation of the entirety of 258 Wing of the RAF. The trapped squadrons just managed to fly out in time to bases further east, after first unleashing hell on the advancing German columns.

1942

Ground crew staff at 268 Squadrons base found themselves left behind and trapped. There were three unserviceable Hurricanes abandoned on the base. Members of the ground crew got two working again in double record time and then found two pilots to fly them. The first took off with the pilot sitting on the lap of one of the ground crew. The second Hurricane then did the same – but with a further member of the ground crew scrunched up in the fuselage radio bay in place of the radio. It was an escape truly worthy of any *Boys' Own* adventure. Fortunately for the remaining ground crew, they managed to get out on a couple of lorries and were escorted to safety by an Allied armoured car company.

The German advance eventually stalled along a line between Gazala and Bir Hakim, partly due to foul weather and their panzers and lorries lacking fuel. Air power now became the dominant factor in the hope to break the stalemate. Hurricane reconnaissance aircraft brought back reports of German supply deliveries and troop movements while Hurricanes flew air cover and relentlessly strafed enemy columns and positions. No 80 Squadron now had Hurricanes capable of carrying a 250lb bomb under each wing and were actually designated as bombers. Fleet Air Arm Hurricanes flew into Egypt to offer further support, while more South African Air Force units arrived in theatre. Significantly, the Hurricane squadrons began to receive the Mk IID which boasted two 40mm Vickers or Rolls Royce cannon and which were especially effective against German vehicles, armoured or otherwise. They soon earned the nickname of *'The flying can openers.'*

Months passed. The Germans were the first to launch another offensive. Tobruk fell again to the Germans on 20 June 1942 and the Allies fell back to a line between El Alamein and the Qattara Depression. By the beginning of September 1942, the Germans were hammering away at this line but could not bring sufficient pressure to bear. Constant (and highly effective) ground attacks by what was now becoming known as 'The Desert Air Force' were ravaging Rommel's essential supply lines. As he paused to draw breath, the Allies launched their own devastating counter-attack out of El Alamein. The attack came at dawn on 23 October 1942. The Hurricane squadrons went on the offensive too, shooting up Axis positions, supply dumps and armour, striking at the Afrika Korps again and again.

✈ L-R: Sea Hurricane re-fuel on HMS *Argus*; RAF preparations in the desert; Hurricanes sitting in wait

✈ Spitfire aboard USS *Wasp*

✈ HMS *Eagle* sinking

By 9 November, German resistance had been shattered and vast lines of enemy trucks jammed the coast road back to Libya. They were attacked again and again from the air, sometimes as part of planned strikes and other times the Hurricanes would be freed up to find and kill 'targets of opportunity'. At the forefront of the strikes were 213 and 238 Squadron Hurricanes, who were actually operating from a makeshift base behind enemy lines codenamed LG125. On Friday, 13 November they succeeded in destroying more than 90 vehicles on the coast road for the loss of two aircraft. The two Hurricane squadrons alternated between hitting the convoys and attacking enemy airfields in the vicinity before being withdrawn from their increasingly unsafe position on 16 November. Four days before, Tobruk fell once more to British and Commonwealth troops

ROCKET MEN

The first Hurricane armed with rockets first flew in February 1942, carrying three rockets under each wing. This would later be increased to four and the rockets would be carried in addition to regular mgs,

⊦ L-R: US Bell P-39 Airacobra; Loading rockets to a Hurricane
⊦ Hurricane scrutinisation in the Soviet Union

anti-tank guns and cannon. Rockets could be fired in pairs, or else as a single salvo of all eight.

Rocket-firing Hurricanes would be used against shipping, road transports, steam locos, barges, airfields, V-1 rocket sites and more from 1942 to 1944, their role being reduced as more and more Hurricane squadrons converted to newer aircraft.

FROZEN OUT IN THE SOVIET UNION

When the British left, their Soviet counterparts fought German bomber fleets on a sporadic basis until the Arctic Winter made flying all but impossible. In just weeks, they still managed to rack up an impressive tally of 'Kills'. As winter reduced in ferocity in early 1942, the Germans came again, this time in much greater numbers. The Hurricanes acquitted themselves well, particularly against Stuka dive bombers but by the Summer of 1942, most of the Hurricanes had been replaced by American warplanes like the P-39 Airacobra or the P-40 Warhawk.

Stalin meanwhile had revised his opinion of the Hurricane. It was *'shit'*, he decided and now the Soviets started demanding Spitfires instead – and many, many of them. Many Soviet pilots agreed with him, criticising both the Hurricane's manoeuvrability and speed. They hated its engine, its reliability, its weaponry, its rate of climb. They just about hated everything. They liked the UHF radio – a bit – but that was it. Fundamental changes to the Hurricane's guns and armour were also soon being made on every air base. Out went the traditional .303 Brownings to be replaced by two 12.7mm Berezin UB machine guns and two 20mm ShVAK cannon. The original armour, which one Soviet pilot claimed could be pierced with a walking stick was also heavily upgraded. In later times, the Soviets would even attach their home-made rockets to the Hurricanes' wings.

'It was a piece of junk rather than a fighter!'
– Soviet pilot Vitaly Klimenko

'(The Hurricane) was very unimportant. Its armament was weak – 12 small-calibre machine guns. This was, for modern German fighters at the time, like peas that could not cause any damage.'
– Soviet pilot Evgeniy Pavlovich Pesterov

One Soviet Commissar, addressing pilots gave a more even-handed assessment;

'The aircraft is fine; it's metal, so it won't catch fire. You can shoot from it. But instead of manoeuvrability and speed – you'll have to use your Russian wits!'

When the Spitfires arrived, they too were sneered at.

Comparatively little is known about the Hurricanes' deployment on the Eastern Front. It was communist policy to concentrate on the achievements of the Soviets and to downplay the role of any foreign help. After all, this was to be 'The Great Patriotic War.' Proper records of the Hurricane in action were not just neglected but actively discouraged.

Despite their unpopularity, Hurricanes were still used far and wide across the Western Soviet Union. Indeed, the Soviet appetite for ever more aircraft was ravenous. Allied Arctic convoys kept bringing them in. Hurricane shipments only finally ceased in 1944, by which time some 2,952 Hurricanes had been sent to the Eastern Front. The vast majority were either Mk IIB's or C's (1,557 and 1009 respectively).

It's an oft-told story that, at war's end, Stalin ordered the surviving Hurricanes to be pushed down mine shafts, so that his people would not see evidence of the assistance the Capitalists had provided. Another version of the story claims that Stalin hid the aircraft simply because he didn't want to pay for them.

TIPPING THE BALANCE

HAWKER HURRICANE THE HISTORY OF A LEGEND

OPERATION TORCH

In November 1942, the Allies launched Operation Torch – the invasion of Algeria and Morocco. Air cover for the operation was partly provided by Fleet Air Arm Sea Hurricanes. After capturing a number of enemy airfields, further RAF Hurricanes were flown in via Gibraltar – sometimes within an hour of the airbase being secured. Six Squadrons in all were assigned – 32, 43, 87, 225, 241 and 253. Hurricane fighters now provided vital air cover for ground operations, whether fending off German bomber formation or providing close air support by strafing and bombing ground positions and columns. The Hurricanes proved effective for a while, but with the arrival of the German Fw-190 (which out-classed the Hurricane in every respect), squadrons in theatre rapidly started to convert from flying Hurricanes to new marks of Spitfires.

Tunisia fell on 13 May 1943, leaving all of Africa in Allied hands. The Hurricane Squadrons charged with guarding North Africa inevitably began to convert to Spitfires. By mid-1944, the Hurricane had effectively vanished from the African continent.

Just one squadron – No 6 – retained its Hurricanes, and they left for Italy in February 1944. The aircraft – adapted to fire rockets – continued to be an effective fighting force over the Adriatic and into Albania and – on one notable occasion – sank a 5,000 ton ship in the Adriatic after blasting it with sixteen rockets. Later in 1944, they switched their attention to Yugoslavia where they proved air support to Greek and Yugoslavian patriot forces. On 7 July 1945, No 6 took up a posting in Palestine and finally surrendered its Hurricanes when in converted to Hawker Tempests on Cyprus in the final days of 1946.

BURMA IN THE BALANCE

It was in the Far East that the Hurricane continued to prove itself despite some squadrons converting to later mark Spitfires to take up the fighter-interceptor role to which the Hurricane was no longer suited. Two Blenheim squadrons actually converted to Hurricanes in August 1943. That same month, the Indian Air Force established its first Hurricane squadron. Three more would follow within a year.

Into 1943 in Burma, Hurricane IIBs began to be replaced by new Mk IICs, with their four 20mm cannon. In the summer of 1943, 20 Squadron received its first Mk IIDs, all fitted with 40mm cannon which could make mincemeat of the poorly-armoured Japanese tanks. Attacks on each other's air bases continued with frequency and ferocity during May 1943. At the same time, Japanese air power was

diminishing, the bulk of their resources being devoted to the air war over the Pacific Islands where the Americans were driving ever-northwards.

In March 1944, when the Japanese advanced on Kohima and Imphal to establish forward bases for the planned invasion of India, there were still eleven Hurricane squadrons in the region to fight them. Often, they operated out of the roughest of jungle airstrips. Rations were worse than basic and malaria and dysentery rife. Their use was mixed. Tac R variants flew reconnaissance. While some fighters flew defensively to protect their air bases, others hunted Japanese supply convoys by night. The Hurricane remained an excellent ground attack aircraft. Apparently, Japanese lorry drivers always insisted in keeping their headlights on and could be spotted from the air with total ease. Consequently, Japanese supply lines suffered considerably. Two more Hurricane squadrons joined the fray later in March and Hurricane operations rose to some 6,000 sorties every month By May, the Japanese had been forced out of Kohima and Spitfires had assumed the bulk of the air defence activities, while Hurricanes concentrated on low level ground attacks for as much as 80% of their sorties.

With Kohima lost, the Japanese threw everything they had into retaining their positions at Imphal but could only hold on to it until July. Then they fell back, at first in an orderly retreat which swiftly turned into chaos, their troops wracked with disease and hunger from lack of supplies. They had lost almost 60,000 soldiers in the fight for Kohima and Imphal and it had broken the back of their army. They fled with next to no air cover and the Allies took full advantage.

Now the RAF Hurricane bombers of 34, 42 and 113 squadrons demonstrated their mastery to co-operate and co-ordinate with Allied army forces, providing vital close air support. They developed the tactic of dropping a bomb on the road ahead of a retreating Japanese convoy and then another to its rear, effectively trapping all the vehicles. They were regularly called upon to drop their bombs on the Japanese who might be less than 600 feet from Allied positions. *'We served as artillery for the army,'* recalled Squadron Leader Jack Rose. The Hurricane bombers sometimes relied on aerial photographs of a target, usually taken by Hurricane Tac R's. When that wasn't possible, Allied army units in forward positions would 'pop smoke' to indicate Japanese positions. After the Hurribombers had gone in, Allied

✈ L-R: Operation Torch in progress; Spitfires being assembled for Operation Torch; Indian pilots were trained to fly Hurricanes

✈ Hurricanes of 6 Squadron ready for take off in Tunisia

army forces would then assault and overrun the stricken Japanese positions while they were still trying to recover. On at least one occasion, Hurribombers were guided onto their target by a string of bright orange golf umbrellas. These were opened up by troops on the ground in a pattern to denote the location of Japanese concealed bunkers.

still prove vulnerable to machine gun and small arms fire from the Japanese. Being shot down remained a real concern, because a pilot's chances of surviving both the Jungle and the Japanese soldier were known to be very slim. Of the 176 Allied pilots who either bailed out or made a crash landing, only ten survived.

By March 1945, Mandalay had fallen to the Allies. Hurricanes moved to new bases close to the city to help the army retake Fort Dufferin, a major Japanese

This formed the pattern of fighting for almost the whole of 1944 and into 1945, the Japanese in headlong retreat eastwards across Burma with the army calling in co-ordinated air strikes from Hurricanes to wipe out their rearguard. The Army often responded with simple messages of thanks that came to be known as *'Strawberries'* (the opposite of *'raspberries'*). There was effectively no Japanese fighters left to worry about and the Japanese had virtually no anti-aircraft guns, but the Hurricanes – flying at such low levels – could

supply centre and troop barracks. They were used to blow holes in the position's earthwork defences, allowing Allied troops to storm inside. The surviving Japanese fled by means of the sewers. 3 May saw the Allies surge into Rangoon. Pockets of fierce Japanese resistance remained, but most were quashed by

TIPPING THE BALANCE

July 1945. The last remaining Japanese positions surrendered on 14 August 1945.

By the end of the campaign in Burma, Hurribombers had dropped well over five million pounds of bombs on Japanese targets. By the end of the campaign they had also been adapted to drop napalm on the enemy. With the collapse of the Japanese Air Force in Burma, the Spitfire interceptors had no-one to fight with and a number were adapted to carry a single 250lb, joining the Hurricanes on ground attack missions for the rest of the war.

All the while, Hurricane squadrons were diminishing in theatre. Apart from those flown by the Indian Air Force, only seven Hurricane squadrons remained by March 1945, the rest having converted to Spitfires or Thunderbolts. By VJ Day – 15 August 1945, just three RAF Hurricane squadrons remained in the whole of the Far East theatre.

✝ A majestic lone low level Hurricane flies alongside the Aya Bridge over the river Irrawaddy in Burma

THE LAST OF THE MANY

A NEW ERA

The end of the Second World War happily coincided with the coming of the jet age and the RAF was almost indecent in its hurry to be rid of the Hurricane and to move onwards and upwards. A large scale flypast was held over London in September 1945 to mark the fifth anniversary of the Battle of Britain. Not one Hurricane was invited to fly. There were no RAF squadrons flying Hurricanes after 1946 and few in positions of authority were sorry to see it go. It had outstayed its welcome and to look at it was to be reminded of the RAF of a different era, one could almost see the ghost of a missing top wing...

14,553 Hurricanes were built in total. Today, it is estimated that less than 15 are still capable of flight.

THE FINAL HURRICANE

The last Hurricane to come off the Langley production line – on 12 August 1944 – was designated PZ865. A IIC variant, it was purchased by Hawker itself and named *'The Last of the Many'*. It even had its name painted on its sides.

THE LAST OF THE MANY

'The Last of the Many' flew in a salute to Hurricanes past before crowds assembled at Langley. Sydney Camm took the salute, with original test pilot George Bulman at the controls. Coverage in both The Times and Flight magazine was respectfully polite but the event very much signified the end of an era.

Post-war, PZ865 was converted for civil use, repainted blue and gold, re-designated G-AMAU and allowed to spend its twilight years on the air racing circuit rejoined Hawker as a target tug for the Harrier project in the 1960s before making a cameo appearance in the feature film 1969 Battle of Britain.

In 1972, PZ865 was given to the RAF's Battle of Britain Memorial Flight, with whom it has proudly flown ever since.

(The Flight also possesses a second Hurricane, LF363, which was also built in 1944 and which served with 309 [Polish] Squadron, the last Hurricane squadron operating in Britain. LF363 co-starred with PZ865 in the film *'Battle of Britain.'* It flew in tribute at Sir Winston Churchill's funeral and at the closure of RAF Bentley Priory. In September 1991, it crashed and burned during rehearsals for a Battle of Britain flypast. Incredibly, it was restored to life by Historic Flying Ltd. The restoration took three years and LF363 is now flying with the BOBMF once more, painted in the colours of 242 Squadron Leader Douglas Bader.)

✈ Battle of Britain Memorial Flight featuring a Hurricane, Lancaster Bomber and Spitfire

✈ (Next pages) Hurricane LF363

HAWKER HURRICANE THE HISTORY OF A LEGEND

THE LAST OF THE MANY